Hypnofacts 9

Trevor Eddolls

This book is dedicated to

Jill, Katy, Harry, Freddy, Phoebe, Jennifer, Andy, Jake, and Rory

First published in 2021

By iTech-Ed Hypnotherapy

16 Brinkworth Close

Chippenham

Wilts SN14 0TL

UK

Typeset by iTech-Ed Ltd

The right of Trevor Eddolls to be identified as the author of this work

has been asserted in accordance with Section 77 of

The Copyright, Designs, and Patents Act 1988

978-1-304-36608-5

Contents

Introduction

Like its predecessors, this book also contains various articles for hypnotherapists covering practical issues such as working with clients with needle phobia, insomnia, and loneliness. There's a section that looks at how successful online therapy can be. There are some thoughts about diet and depression, the positive and negative side of boredom, and the effects of kindness. And there are more theoretical issues around genetics, positive psychology, and ideomotor responses.

Again, the articles assume a model of the brain in which core activities (such as telling the heart to beat) are handled by the brain stem; more protective functions (such as fighting, fleeing, feeding, and reproductive behaviour) are handled by the primitive emotional brain; and higher functions (such as problem solving, maintaining attention, and controlling emotional impulses from the primitive brain) are handled by the intellectual brain. In terms of physical parts of the brain, these three areas more-or-less match up to the brain stem and cerebellum, the limbic system, and the cerebral cortex. It also assumes that the primitive emotional brain is very fast and the intellectual brain is much slower and tends to be used less.

In addition, the book assumes that the mind and body make up a single functioning system that is affected by each component and the environment they are in.

And it assumes a solution-focused model for hypnotherapy – moving clients towards their desired goals rather than worrying about the problem itself and its origin.

Bring me sunshine

A look at the benefits of optimism.

So, you're looking at a glass – is it half empty or is it half full? Or is it just a really nice glass?

Optimism is all about being hopeful and confident about the future or the success of something. But, is that the best way to approach life? Let's see what research suggests.

It appears that optimism is good for your heart. There's a story that on 12 July 1988 France beat Brazil in the World Cup final, and on that day, there were fewer cardiovascular deaths in France than the average for the previous week. And that's been attributed to a burst of optimism!

M F Scheier *et al* (1999) studied 309 middle-aged patients scheduled for coronary bypass surgery. They found that compared with pessimistic people, optimistic people were significantly less likely to be rehospitalized for a broad range of problems, such as postsurgical sternal wound infection, angina, myocardial infarction, and the need for another bypass surgery. They concluded that optimism predicts a lower rate of rehospitalization after coronary artery bypass graft surgery. And fostering positive expectations may promote better recovery.

Similarly, optimism is good for people with high blood pressure. A Finnish study of 616 middle-aged men with normal blood pressures when the study began had their mental outlook evaluated with questions about their expectations for the future. They were also evaluated for cardiovascular risk factors such as smoking, obesity, physical inactivity, alcohol abuse, and a family history of hypertension. Over a four-year period, highly pessimistic men were three times more likely to develop high blood pressure than optimists, even after other risk factors were taken into account.

A US study of 1,306 men with an average age of 61 evaluated them for an optimistic or pessimistic outlook, as well as for blood pressure, cholesterol, obesity, smoking, alcohol use, and family history of heart disease. None of them had a diagnosis of coronary artery disease when the study began. Over the next 10 years, the most pessimistic men were more than twice as likely to develop heart disease than the most optimistic men, even after taking other risk factors into account.

Looking at overall health, a study of 2,300 older adults (over 85 years old) over a 2-year period found that people with a positive outlook were much more likely to stay healthy and enjoy independent living than less cheerful people. Another study of 447 patients over a 30-year period found optimism was linked to a better outcome on eight measures of physical and mental function and health.

Optimists also live longer. A US study of 839 people concluded, after a 30-year follow-up, that optimism was linked to longevity. They suggested that for every 10-point increase in pessimism on an optimism-pessimism test, the mortality rate rose by 19 percent.

Another US study of 6,959 students found that over the next 40 years, 476 of the people died from a variety of causes, with cancer being the most common. They concluded that pessimists had a 42 percent higher rate of death than the most optimistic students.

A Dutch study of 941 people aged between 65 and 85 found that people who demonstrated dispositional optimism at the start of the study enjoyed a 45 percent lower risk of death during a nine-year follow-up period.

Why do optimists live longer? A 2008 study of 2,873 healthy people found that a positive outlook on life was linked to lower levels of cortisol, even after taking age, employment, income, ethnicity, obesity, smoking, and depression into account. Other possible benefits include reduced levels of adrenalin, improved immune function, and less active clotting systems.

Optimists also seem to make healthier choices. A study by Steptoe *et al* (2006) of people aged between 65 and 80 years found that optimism was correlated with healthy behaviours such as abstaining from smoking, moderate consumption of alcohol, the habit of walking briskly, and regular physical activity, regardless of demographical factors, current psycho-physical conditions, and body mass.

A study by Lee *et al* (2019) confirmed that people with greater optimism tend to live longer than pessimists, on average. This decades-long study identified a strong correlation between optimism and exceptional longevity, which is described as living to age 85 or older.

It's also good for any athletes you work with to be optimistic. An optimistic athlete is more likely to be persistent and committed during the action phase of working towards a goal and are more likely to be able to tolerate uncontrollable suffering (according to Espahbodi *et al,* 1991). Optimistic athletes believe that successful performance is within their control, and the reason is unchanging, ie they are a good player. They view an unsuccessful performance as a temporary setback that's caused by something out of their control, eg bad weather. This means that their self-esteem isn't impacted because they believe that they are in control of their performance level and not of the negative weather, for example. And this, in turn, leads to optimism about future positive performances.

Gordan & Kane (2001) found optimism led to better performances and less variability. An older study by Carver *et al* (1979) found that optimism helped athletes to overcome adversities, was motivating, and increased persistence. And Rettew & Reivich (1995) found optimism led to more wins.

We're perhaps more interested in what's going on inside people's brains. E C Chang *et al* (2001) and S L Hart (2008) found an inverse correlation between optimism and depressive symptoms. And J K Hirsch *et al* (2007) found the same inverse correlation between optimism and suicidal ideation.

M F Scheier *et al* (1986) found a significant positive relationship between optimism and different aspects of life, such as coping strategies focalized on the problem, looking for social support, and emphasis of the positive aspects of the stressful situation.

Other research (L S Nes *et al* 2006) confirms that optimism is positively correlated with those coping strategies thought to eliminate, reduce, or manage the stressors and negatively correlated with those employed to ignore, avoid, or distance oneself from stressors and emotions.

These all go to show that helping our clients to be more optimistic is going to help them with the usual ups and downs of life.

Dr Richard Davidson, in research using functional MRI scanners and advanced EEG analysis, found that when people are emotionally distressed (anxious, angry, depressed) the most active sites in the brain are circuitry converging on the amygdala and the right prefrontal cortex. We probably knew this. And when people are in a positive mood (upbeat, enthusiastic, and energized) there's heightened activity in the left prefrontal cortex. Dr Davidson also found that volunteers with more left-side activity who watched amusing films had a far stronger pleasant response, while those with more right-side brain activity who watched distressing films had far stronger negative feelings.

So, positive moods are associated with more left-brain activity, and negative moods with more right-brain activity. The good news is that by consciously altering their thought processes, our clients can literally re-wire their brain.

So, research has shown that optimism is correlated with many positive life outcomes including increased life expectancy, general health, better mental health, increased success in sports and work, greater recovery rates from heart operations, and better coping strategies when faced with adversity.

But where does optimism come from? Can it be learned? A study by Robert Plomin *et al* (1992) suggested that optimism is about 25 percent inheritable. The other 75 percent comes from other factors such as socioeconomic status that are probably out of our control.

How can we help our clients to be more optimistic? Dr Davidson taught mindfulness to workers in high-stress jobs who were right-prefrontal cortex users. After two months of training (for three hours each week), they moved to the left prefrontal cortex and reported feeling less anxious, more energized, and happier.

Some other techniques you can try with clients include:

- Look for the good – instead of highlighting the bad events, make a note of the good things. That's what we ask clients – "what's been good". We want them to start noticing those sparkling moments. This may well train their brain to think more positively.

- Positive interactions – too often conversations with others become moaning session or gossip fests. We need to encourage clients to be positive and mix with positive people. That will make them feel happier and more optimistic.

- Bucket emptying – to get people out of their right prefrontal cortex, we need to help them to empty their metaphorical emotional buckets. And we can do that in the usual way of helping them to relax, helping them to get a good night's sleep, and building their confidence.

- Turn off the news – your thoughts become your reality. If a client is continually thinking about all the bad news, it will lead them to have a negative and rather bleak view of the world. This will move them into their right prefrontal cortex and make them less optimistic. So, they should stop watching the news and searching the Internet for news stories (or at least cut down to the bare minimum).

Psychologist Julie Norem's research suggests that the phenomenon of defensive pessimism, the cognitive strategy of setting low expectations and considering worst-case scenarios of future events (ie not getting your hopes high), can help with managing anxiety and gaining a sense of control. Defensive pessimism is most helpful when potential negative outcomes are important and there are things you can do to prevent these outcomes.

- Keep a gratitude journal – towards the end of each day, ask your client to write down a couple of things or events that happened during the day that they feel grateful for. This will get them to focus on the positives of their day and cultivate an optimistic mindset.

- Recognize the things they can control and the things they can't – and don't spend time worrying or ruminating about the things they can't control. Mindfulness has a technique to stop people ruminating over the things that they have found stressful during the day. People allow thoughts to enter their brain, then gently push them away without judgement.

- Recognize their own talents – too often people don't see the strengths that they have used during the day, or don't remember the nice things that people have said to them during the day. To be optimistic, it's important for a person to acknowledge what they have done and when people have appreciated them.

One well-known cognitive bias is the 'optimism bias'. This causes people to believe that they are less likely to experience a negative event. For example, they think they're less likely to catch Covid-19 than anyone else.

Does optimism or positive thinking ever have a downside? It seems the answer is sometimes. It's been suggested that negative emotions can motivate people to change things for the better for themselves and for others. And feeling a wide range of positive and negative emotions helps a person to find meaning in life and grow as a person.

Being over-optimistic can result in people miscalculating risks and making bad decisions. If a person is doing a dangerous sport, and they assume a risky technique will have a positive outcome for them, that can have disastrous consequences.

Optimism can also lead to complacency. You hear people say, "it'll be alright". Visualizing achieving something (eg getting a new job or climbing a mountain) can result in a drop in energy levels, resulting in a worse performance during the actual event. Apparently, extreme optimists are less likely to clear outstanding credit card balances, save less money, and work fewer hours than even other optimists.

Bearing those issues in mind, we can still conclude that being optimistic is generally good for our clients' health, helps them to live longer, and can be better for their sporting performance. We can see that optimism can be learned and we can help clients to become less pessimistic and more optimistic – and enjoy those benefits of optimism. But, of course, being optimistic is not all good, and we should be aware of the dangers that come for clients with an over-optimistic outlook.

References:

https://believeperform.com/optimism-in-sport/

https://www.nytimes.com/2003/02/04/health/behavior-finding-happiness-cajole-your-brain-to-lean-to-the-left.html?mcubz=1

https://www.hongkiat.com/blog/optimism-positive-thinking/

https://www.ncbi.nlm.nih.gov/pmc/articles/PMC2894461/

https://www.health.harvard.edu/heart-health/optimism-and-your-health

https://www.psychologytoday.com/gb/blog/the-athletes-way/201908/optimism-study-gives-optimists-more-reason-be-optimistic

http://positivepsychology.org.uk/optimism-pessimism-theory/

https://www.nbcnews.com/better/health/how-train-your-brain-be-more-optimistic-ncna795231

https://www.psychologytoday.com/us/blog/between-cultures/202104/7-myths-about-optimism-and-pessimism

Covid-19 and needle phobia

How solution-focused hypnotherapy can help people scared of injections.

With the arrival of the Pfizer-BioNTech COVID19 vaccine just before Christmas 2020, the UK population divided into three groups. The first group were keen to get the vaccination as soon as possible in order to get on with their lives. The second group were people who just don't like needles and, even though they probably want the vaccination, were not going to have it. And the third group were described as the needle hesitancy or vaccine hesitancy group. These are people who can't decide about having the vaccine, perhaps thinking that it's not safe The World Health Organization (WHO) lists this group among the top 10 threats to global health. The Wellcome Global Monitor survey, looking at 2018 data, found that, in France, one in three people believe that vaccines are not safe. In the Ukraine, just half of those surveyed trusted vaccines. Other surveys found that only 15 percent of people in Russia are willing to get vaccinated as soon as possible. In the USA, the figure is 59 percent. A YouGov survey in the UK found 80 percent of people are willing to have or already have had a coronavirus injection. The number of people against vaccination seems to be growing around the world.

With the Oxford AstraZeneca vaccine now also being given to older residents in a care homes and staff in those homes, as well as people over 80 and frontline health and social care workers, let's look at that second group. A 2003 survey by Nir *et al* entitled "Fear of injections in young adults: prevalence and associations", published in the *American Journal of Tropical Medicine and Hygiene*, found that 3.5 to 10 percent of the general population have needle phobia anxiety disorder.

For those of you who like this kind of thing, trypanophobia is the name given to an extreme fear of medical procedures involving injections or hypodermic needles. You may also hear it called aichmophobia or belonephobia, which really mean fear of sharply-pointed objects.

How do you know if you have needle phobia? According to Anxiety UK, if you can answer YES to the following questions, it is likely that you do. During the last 6 months:

- Have you experienced a marked, persistent, and excessive fear of needles?
- Has exposure to needles almost invariably provoked an immediate anxiety response in you?

It seems that there are four types of needle phobia.

- **Vasovagal** is where people fear the sight, thought, or feeling of needles or needle-like objects. This leads them to faint (vasovagal syncope) because of a drop in blood pressure. The condition starts with momentary high blood pressure and a fast heart rate (a fight-or-flight response) followed by both decreasing enormously at the moment of injection. Very rarely, the drop in blood pressure caused by the vasovagal shock reflex may cause death, but this is usually linked to an underlying condition.

- **Associative** is where a traumatic event causes the person to associate all procedures involving needles with the original negative experience.
- **Resistive** is where a person doesn't just fear needles or injections but also being controlled or restrained.
- **Hyperalgesic** is where people have an inherited hypersensitivity to pain (hyperalgesia). So, the pain of an injection is unbearably great. Usually, some form of anaesthetic helps these sufferers.

And, some people experience more than one kind of needle phobia.

Needle phobia is unusual for a phobia in that it can be a direct cause of death in some documented cases – and probably the cause in many more undocumented cases because of all the people who avoid medical and dental treatment because of the condition.

So, how can solution-focused hypnotherapy help?

Firstly, your hypnotherapist will tell you that a fear of needles is not uncommon and tell you that the people giving the injection will be perfectly used to seeing people with that particular fear. So, they will recommend that you tell the clinician at the beginning that you don't like needles – it's nothing to be embarrassed about. Telling the staff means that they are better placed to help you. They will be able to answer any questions you have and put you at ease.

If you have fainted in the past, the clinicians may suggest that you lie down while they give the injection. I expect that the other techniques suggested by your solution-focused hypnotherapist will mean that doesn't happen this time, but lying down won't do any harm.

For people with associative or resistive types of needle phobia, solution-focused hypnotherapy can help by emptying a person's metaphorical stress bucket – helping them to feel less anxious about some things and more confident about other things. For them, techniques such as rewind – where a person plays a video in their mind of an unpleasant event – really works. And they play the video forwards and backwards, faster and faster, even with added silly music, until all the emotions associated with the event are gone, and they no longer fear it. This works well, and is combined with a reframe – where a person repeatedly imagines how they would like to behave in a situation that they had previously found scary. So that when they next encounter the situation, they behave just how they imagined they would.

Even with hyperalgesic needle phobia, relaxation and bucket emptying help reduce the sensation of pain that a person feels. Emla cream is often used at the surgery to numb the pain of the injection.

But with vasovagal needle phobia, relaxing won't help if a person is going to naturally lower their blood pressure so much that they faint. So, how can solution-focused hypnotherapy help this group?

The most successful technique seems to be the applied tension technique. With this, a person can increase their blood pressure back to normal – so they don't faint. Here's what Guy's and St Thomas' Hospital suggest that people do:

1 Sit down somewhere that's comfortable.

2 Tense the muscles in your arms, upper body, and legs, and hold this tension for 10 to 15 seconds, or until you start to feel the warmth rising in your face.

3 Release the tension and go back to your normal sitting position.

4 After about 20 to 30 seconds, go through the tension procedure again until you feel the warmth in your face.

5 Repeat this sequence five times.

If you can, practise this sequence three times every day for about a week, before being vaccinated.

It's suggested that if a person gets headaches after doing this exercise, they don't tense the muscles in their face and head. Also, people should be careful when tensing any part of their body where they have any health problems.

Although people with vasovagal needle phobia don't want to relax and lower their blood pressure during the vaccination, they do want to be relaxed when travelling to get the vaccination and when waiting to be vaccinated. Your hypnotherapist can help with that with some of the following techniques.

Breathing techniques such as 7-11 breathing or square breathing are ways of breathing slower and relaxing. You can watch your abdomen rise and fall rather than your chest. Your hypnotherapist will show you how to do these techniques.

Your therapist may suggest that you smile! Research by Pressman *et al* published in 2020 in the journal *Emotion* found that smiling could reduce needle pain by 40 percent. They reported that "the Duchenne smile and grimace groups reported approximately 40% less needle pain versus the neutral group".

So, what's a Duchenne smile? It's the one where you not only lift the corners of your mouth but also lift your cheeks and crinkle your eyes at the corners.

Your solution-focused hypnotherapist will also suggest that you use distraction techniques. This is a way of focusing on something else and not keep thinking about the jab! Your hypnotherapist will have given you a download or CD. You can listen to that as a way of relaxing before being called in. Or, you can listen to music that reminds you of dancing like mad in the mosh pit to your favourite band. Or you can choose music that takes you away to distant destinations.

You can remind yourself of a holiday or family party and step through the events that happened in real time. You can picture the scene in vivid colour with everything bright and in focus. You can imagine the events are taking place on the largest cinema screen ever. You can listen again to sounds with crystal clarity. And you can feel again those feelings that you experienced at the time.

Or you can lose yourself in a book. Or you can focus on a game on your phone.

CBT (Cognitive Behavioural Therapy) uses an exposure technique to help people get used to something they are originally fearful of. If you think that this is the sort of thing you like, Anxiety UK has published a self-administered behavioural exposure program in a PDF called "Injection Phobia and Needle Phobia: A brief guide".

There are three steps in the self-administered behavioural exposure hierarchy:

1 Relaxation – which could be by practising progressive muscle relaxation, breathing exercises, or meditation.

2 Constructing an anxiety hierarchy or 'fear ladder' – where a person writes down a list of all of the situations related to needles that they fear, arranged in order of difficulty.

3 Pairing relaxation with the situations detailed in their hierarchy – where a person climbs the ladder (by thinking about or acting out each step) from bottom to top, exposing themself to the fear for a tolerable amount of time before taking time to relax.

Lastly, in an article entitled "How to Give Your vaccine a Boost", the *New Scientist* gave some hints and tips on how to get the most from your inoculation. They suggested that you:

- Try not to stress. Research on the Hepatitis B vaccine shows that stressed people take a bit longer to build up their immune response.

- Do all you can to ensure a that you're in a positive mood on the day of the vaccine.

- Get a good night's sleep for at least two nights before the vaccine. The antibody response is increased when a person sleeps for more than 7 hours.

- Talk to friends because feeling lonely can impact a person's antibody response.

- Hug the people you live with, if you can. Video call friends and family. And avoid 'doom scrolling' through social media.

- Control your alcohol intake because high levels of alcohol consumption can impair the body's immune response, although moderate drinking is unlikely to have much effect.

- Exercise. People with an active lifestyle have higher antibody responses. So, do some exercise before your vaccine in order to activate your immune response. In addition, exercise is also an analgesic. It seems that feelings of pain at the site of the vaccine injection, or feeling tired or unwell afterwards, can be reduced by exercise.

Conclusion

Whatever type of needle phobia you experience, it's worth contacting a solution-focused hypnotherapist to help you overcome your fear and benefit from the vaccination against the Covid-19 virus.

References:

The Wellcome Global Monitor survey: https://wellcome.org/reports/wellcome-global-monitor/2018/chapter-5-attitudes-vaccines

Overcoming your fear of needles: https://www.guysandstthomas.nhs.uk/resources/patient-information/all-patients/overcoming-your-fear-of-needles.pdf

Smile (or grimace) through the pain? The effects of experimentally manipulated facial expressions on needle-injection responses: https://psycnet.apa.org/record/2020-88213-001

Injection Phobia and Needle Phobia: A brief guide: https://www.gmmh.nhs.uk/download.cfm?doc=docm93jijm4n6539.pdf&ver=9064

https://www.newscientist.com/issue/3320/

Insomnia

What it is, why it's bad, and what to do about it.

Insomnia – we've all had it. And many more people seem to be suffering with it during the lockdown than at any time in the recent past. But what is it?

Insomnia is a sleep disorder in which people have trouble sleeping. It's sometimes called sleeplessness. People with insomnia may have difficulty falling asleep, or they may wake up early and can't get back to sleep. They may wake up several times during the night or they may lie awake at night. The trouble with not getting enough sleep during the night is that people feel sleepy the following day, although they find it hard to nap during the day even though they're tired. They also feel low on energy, they can be irritable, and may feel depressed. Episodes of insomnia may be short-term or may go on for much longer.

Sometimes, insomnia is linked to other conditions such as stress, chronic pain, heart failure, hyperthyroidism, heartburn, restless leg syndrome, menopause, certain medications, and drugs such as caffeine, nicotine, and alcohol. Insomnia can also be

> Sleep debt or sleep deficit is the name given to the cumulative effect of not getting enough sleep.

associated with working night shifts and sleep apnoea (where a person has pauses in their breathing or periods of shallow breathing during sleep). Poor sleep quality is defined as the individual not reaching stage 3 (also called delta sleep), which has restorative properties. We'll talk about these stages of sleep later.

Let's look at some statistics. It's been estimated that between 10 and 30 percent of adults have insomnia at any given point in time, and up to half the population has insomnia in a given year. About 6 percent of people have insomnia that is not due to another problem and lasts for more than a month. Women are more often affected than men.

People in the following groups have a higher chance of experiencing insomnia. They:

- Are older than 60
- Have a history of mental health disorder including depression, etc.
- Are experiencing emotional stress or work stress
- Work late/night shifts
- Have travelled through different time zones
- Have chronic diseases such as diabetes, kidney disease, lung disease, Alzheimer's, or heart disease
- Have alcohol or drug use disorders
- Have gastrointestinal reflux disease
- Are heavy smokers.

You know you've got insomnia if you:

- Have difficulty falling asleep, including difficulty finding a comfortable sleeping position.
- Wake during the night and are unable to return to sleep or you wake up early.
- Are not able to focus on daily tasks, and have difficulty in remembering.
- Experience daytime sleepiness, depression, or anxiety.
- Feel tired or have low energy levels during the day.
- Have trouble concentrating.
- Are irritable, or act aggressively or impulsively.

Insomnia can be classified as transient, acute, or chronic.

- Transient insomnia lasts for less than a week. It can be caused by another disorder, by changes in the sleep environment, by the timing of sleep, by severe depression, or by stress. Its consequences (ie sleepiness and impaired psychomotor performance) are similar to those of sleep deprivation.
- Acute insomnia is the inability to consistently sleep well for a period of less than a month. Insomnia is present when there is difficulty initiating or maintaining sleep or when the sleep that is obtained is non-refreshing or of poor quality. These problems occur despite adequate opportunity and circumstances for sleep, and they result in problems with daytime function. Acute insomnia is also called short-term insomnia or stress-related insomnia.
- Chronic insomnia lasts for longer than a month. It may or may not be caused by another disorder. People with high levels of stress hormones or shifts in the levels of cytokines are more likely than others to have chronic insomnia. Its effects can vary according to its causes, but they may include: muscular weariness, hallucinations, and/or mental fatigue. Chronic insomnia can cause double vision.

Insomnia can be measured using the Athens Insomnia Scale (AIS). It assesses eight factors on a scale of 0–3. The factors are: sleep induction; awakenings during the night; final awakening (how early); total sleep duration (how sufficient); sleep quality; wellbeing during the day; functioning capacity during the day; and sleepiness during the day. The higher the number on the scale, the worse things are. A total score of 6 or above equates to a diagnosis of insomnia.

Sleep opportunity is how many hours you need to spend in bed in order to get enough sleep. It's calculated by adding together the number of hours of sleep you need, plus the number of hours you know you'll need to fall asleep, plus how long it takes you to wake up.

It's worth noting that sleep onset insomnia, which is a difficulty falling asleep at the beginning of the night, is often a symptom of anxiety disorders. Delayed sleep phase disorder (DSPD) is often misdiagnosed as insomnia. DSPD is where sleep onset is delayed until much later than normal while waking up is also much later in the day.

Paradoxical insomnia (previously called sleep state misperception) is where people say they have slept poorly, or not slept at all, but they actually have. Using electrodes or other sleep monitoring devices, it can be shown that there's a huge mismatch – the person has slept much better than they report. People with paradoxical insomnia have the illusion of poor sleep, when it's not actually poor.

> Sleep pressure is simply the pressure for us to go to sleep. It becomes greater the longer we have been awake.

So, what are the causes of insomnia? There are two models of why people get insomnia. In the cognitive model, people are thought to be unable to get to sleep through rumination or hyperarousal. The second model, the physiological model, is based on what's found in people with insomnia. Firstly, there's increased urinary cortisol and catecholamines (eg adrenalin and noradrenalin), which have been found suggesting increased activity of the HPA axis and arousal. Secondly, people with insomnia have been found to have increased global cerebral glucose utilization during wakefulness and NREM sleep. Thirdly, increased full body metabolism and heart rate has been found in people with insomnia. These findings suggest that there's a dysregulation of the arousal system, cognitive system, and HPA axis that contribute to insomnia. What's not known is which is the 'cause' and which is the 'effect'.

Other common causes of insomnia include: noise, a room that's too hot or cold, an uncomfortable bed, alcohol, caffeine or nicotine, recreational drugs like cocaine or ecstasy, jet lag, or shift work.

As mentioned earlier, sleep disturbance is about twice as common in women as men. And around half of post-menopausal women experience sleep disturbances. This seems to be due in part to changes in hormone levels. Also, changes in sex hormones in both men and women as they age may account in part for increased prevalence of sleep disorders in older people.

> The grogginess people sometimes feel when they first wake up is called sleep inertia.

So, what does ordinary sleep look like for those people who get it? Sleep is divided into four stages (it used to be five, but stages 3 and 4 were combined). The first three stages are non-REM (NREM) sleep, and the fourth stage is REM (Rapid Eye Movement) sleep. These stages are:

- Stage 1 – which can be considered a transition period between wakefulness and sleep, and lasts around five to 10 minutes. The brain produces high amplitude theta waves, which are very slow brain waves.

- Stage 2 – where people become less aware of their surroundings, their body temperature drops, and their breathing and heart rate become more regular. It lasts around 20 minutes. The brain begins to produce bursts of rapid, rhythmic brain wave activity called sleep spindles. People spend around 50 percent of their total sleep in this stage.

- Stage 3 delta sleep – where a person's muscles relax, and blood pressure and breathing rate drop. Slow brain waves (delta waves) are generated. People become less responsive, and noises and activity in the environment may not get a response.
- Stage 4 REM sleep – where the brain becomes more active, and the body becomes relaxed and voluntary muscles become immobilized. Dreams occur and the eyes move rapidly. There is also an increase in respiration rate, and increased brain activity. People spend around 20 percent of their total sleep in this stage.

Sleep begins in stage 1 and progresses into stages 2 and 3. After stage 3, stage 2 sleep is repeated before entering REM sleep (stage 4). Once REM sleep is over, the body usually returns to stage 2 sleep. Sleep cycles through these stages approximately four or five times throughout the night. Typically, sleep cycles last around 90 minutes. The first cycle of REM sleep might last only a short amount of time, but each cycle becomes longer through the night. REM sleep can last up to an hour as sleep progresses.

What's going on inside your brain when you go to bed? In good sleepers, the amygdala, the hippocampus, and the alertness regions of the brain stem become less active as they begin to fall asleep. With insomniacs, these regions stay active. Their thalamus also stays active. People with insomnia also have lower quality of sleep with shallower and less powerful brainwaves during NREM sleep, and more fragmented REM sleep.

The hypothalamus contains the suprachiasmatic nucleus (SCN), which receives information about light exposure from the eyes and controls a person's circadian rhythms. It produces the neurotransmitter gamma-aminobutyric acid (GABA), which reduces the activity of arousal centres in the hypothalamus.

The brain stem communicates with the hypothalamus to control the transitions between wake and sleep. The brain stem also produces GABA, which reduces the activity of arousal centres in the brain stem.

During most stages of sleep, the thalamus becomes quiet, ignoring messages from the outside world. But during REM sleep, the thalamus is active, sending the cortex images, sounds, and other sensations.

The pineal gland receives signals from the SCN and increases production of the hormone melatonin, which helps put a person to sleep once the lights go out. When the eyes receive light from the sun, the pineal gland's production of melatonin is inhibited. When the eyes do not receive light, melatonin is produced in the pineal gland and a person feels tired.

The basal forebrain promotes sleep and wakefulness, while part of the midbrain acts as an arousal system. Adenosine is produced by astrocytes (a type of glial cell) in the basal forebrain. Adenosine is a neurotransmitter/neuromodulator affecting the sleep process, particularly the initiation of sleep. In the brain, it is an inhibitory neurotransmitter and inhibits many processes associated with wakefulness. While awake, levels of adenosine in the brain continue to rise, increasing a person's level of sleepiness. Adenosine levels decrease during sleep.

Neurons, located predominantly in the hypothalamus, produce orexin, which is a neuropeptide that seems to promote wakefulness. It also regulates arousal, feeding, energy expenditure, and modulates visceral function. The role of the orexin system is to integrate metabolic, circadian, and sleep debt influences to determine whether an animal should be asleep or awake and active. Orexin neurons strongly excite various brain nuclei with important roles in wakefulness including the dopamine, noradrenalin, histamine, and acetylcholine systems, and appear to play an important role in stabilizing wakefulness and sleep.

> Your immune system can make you feel sleepy when it's fighting an infection. Lack of sleep can impair your immune system.

If you get enough sleep, it:

- Reduces stress
- Reduces the risk of depression
- Makes you more alert
- Improves your memory
- Cleans up your brain
- Makes you cleverer
- Helps your body repair itself
- Reduces inflammation
- Keeps your heart healthy
- May prevent cancer
- May help you lose weight.

But what happens if you don't get enough sleep? Insomnia:

- Is linked with depression and anxiety
- Makes you forgetful
- Impairs your judgement
- Cognitively impairs your thinking
- Causes accidents, eg falling asleep at the wheel
- Is linked to health issues (heart attack, stroke, diabetes)
- Kills sex drive
- Ages your skin
- Can cause weight gain
- Increases the risk of death.

So, if your client does have insomnia, what can you do to help? Milton Erickson had a client with insomnia. He famously used an avoidance-avoidance bind to resolve his client's symptoms. The client was a meticulous elderly man who prided himself on doing all his own housework. All of it except waxing the floors, which he hated. Erickson told the man that there was an obvious solution to his insomnia problem, but he might not like it (a typical Erickson comment!). The man insisted that he would do whatever was necessary in order to sleep. Erickson, as usual, was reluctant to tell the man what he needed to do, and the man insisted that he'd do whatever was necessary, giving various examples of how persistent he was in dealing with difficult problems. Erickson finally told him that if he wasn't asleep within fifteen minutes of going to bed, he had to get up and wax floors until he felt he could sleep. After that, if he was still not asleep within fifteen minutes, he had to get up again and continue this procedure until he was asleep. The end result was that the man had well-waxed floors and slept very well.

Clients may want to try medications from their doctor, but, remember, these are not recommended for more than four or five weeks, and chronic insomnia can last longer than that.

Therapists can also recommend (what's called) sleep hygiene changes. Sleep hygiene is a common term for all of the behaviours relating to the promotion of good sleep. So, what should you advise? Here are some ideas. Clients should:

- Go to bed and wake up at the same time every day (including weekends)
- Make sure their mattress, pillows, and covers are comfortable
- Find soothing ways to relax into sleep, including use of white noise
- Use the bed only for sleeping (and sex)
- Make sure the bedroom is dark, cool (18°C), and quiet
- Relax at least 1 hour before bed
- Take a bath or read a book before bed
- Eat a small amount of food rich in carbohydrates, such as cereal. In addition, the milk with the cereal contains tryptophan which promotes the production of melatonin.
- Exercise regularly during the day. Exercise not only improves sleep, but good sleep improves a person's performance of exercise.
- Get exposure to sunlight during the day
- Sprinkle lavender or neroli oil on their pillow before sleeping.
- Try some moderately difficult mental arithmetic
- Get out of bed after 15 minutes of wakefulness and do something relaxing or non-stimulating for twenty minutes before going back to bed.
- If in these days of working from home (#WfH) where their bedroom is also their office, make the room look different during the daytime and the nighttime, eg put a different blanket over the bed during the day.

And here are some things for clients to avoid:

- Do not sleep-in after a bad night's sleep
- Do not nap during the day
- Do not strengthen any links between bed and the idea of not sleeping.
- Do not keep checking the time
- Do not watch television or use devices with screens right before going to bed
- Do not smoke or drink alcohol, tea, or coffee at least 2 hours before going to bed
- Do not eat a big meal late at night
- Do not exercise at least 2 hours before bed.

One technique, that is sometimes used, is to restrict the amount of time that people with insomnia spend in bed – maybe to just six hours. Keeping people awake for longer builds up the sleep pressure, and people fall asleep faster and sleep better. It also increases their expectation that they will sleep better in future.

You can also recommend that clients who can't sleep try 'paradoxical intention'. This is where, instead of lying there struggling to get to sleep, a person remains passively awake and avoids any efforts to fall asleep. This eliminates any performance anxiety that may inhibit sleep onset. People often naturally fall asleep.

If your client's insomnia is caused by stress, anxiety, or feeling depressed, we can help with those conditions. It's the usual bucket emptying and emphasis on the 3Ps – positive thoughts, positive actions, and positive interactions. (And that fourth P – purpose.) And, of course, listening to your download/CD will help them to get to relax and get to sleep – whether that's when they first go to bed or in the middle of the night.

Sleep well.

References:

https://en.wikipedia.org/wiki/Sleep_apnea

https://www.nhs.uk/conditions/insomnia/#overview

https://www.hypnotherapy-directory.org.uk/memberarticles/twelve-steps-to-better-sleep

https://www.ninds.nih.gov/Disorders/Patient-Caregiver-Education/Understanding-Sleep

Matthew Walker. Why We Sleep: The New Science of Sleep and Dreams. Penguin. ISBN-10: 9780141983769

https://www.verywellhealth.com/the-four-stages-of-sleep-2795920

Nightmare clients

How to help clients experiencing nightmares.

Children and adolescents experience more nightmares than adults. Although, half of adults experience nightmares sometimes, and nightmares are more usual among women than men. Nightmares can increase with traumatic or adverse events, irregular sleep, sleep deprivation, and jet lag.

The first question to answer is whether hypnotherapy can help someone with nightmares. And the research shows that the answer is 'yes'. The anecdotal evidence also shows that the answer is a resounding yes.

A 2007 study by Haun *et al* published in the *Journal of Clinical Sleep Medicine* found that one or two sessions of hypnotherapy might be an efficient first-line therapy for patients with certain types of parasomnias. Parasomnias are undesirable events or experiences that occur either during sleep or within close proximity to sleep and include nightmares, sleepwalking, etc.

Other studies have also found hypnotherapy to be helpful for people experiencing nightmares, eg Gerard Kennedy 2002.

Are nightmares a bad thing? A 2019 study published in the journal *Human Brain Mapping*, demonstrates a strong link between the emotions we feel in both sleep and wakefulness. They concluded that when we wake up from a bad dream, the brain regions linked to emotional control tend to respond to fear-inducing situations much more effectively. The study identified two brain regions implicated in the induction of fear experienced during the nightmare: the insula and the cingulate cortex. The insula is also involved in evaluating emotions when we're awake, and is automatically activated when someone feels afraid. The cingulate cortex plays a role in preparing motor and behavioural reactions in the event of a threat.

The study found that the longer someone felt fear in their dreams, the less the insula, cingulate, and amygdala were activated when the same person looked at negative pictures. They also found the activity in the medial prefrontal cortex – which is known to inhibit the amygdala in the event of fear – increased in proportion to the number of frightening dreams! The findings demonstrate a very strong link between the emotions we feel in both sleep and wakefulness and reinforce a neuroscientific theory about dreams – that we simulate frightening situations in our dreams to better react to them once we're awake!

However, researchers did wonder whether, when a certain threshold of fear is exceeded in a dream, it loses its beneficial role as an emotional regulator.

A 2009 study by J Roberts *et al* analysed the dreams and stress levels of 624 high school students and found that those who reported being distressed by their dreams were even more likely to report suffering from general anxiety. The study didn't find whether the nightmares made the children more stressed or whether it prevented them being even more stressed. However, a small 2019 study by Louis-Philippe Marquis *at al* found that when people who frequently recalled their nightmares were shown disturbing images, the areas of the brain associated with negative emotions in dreams were activated, demonstrating that nightmares could actually enhance waking-life distress.

Canada's Tore Nielsen's team found that theta brainwave activity is higher in people who have frequent nightmares. It might be that such people put more brain effort into processing their emotions and experiences during dreaming. Mark Blagrove and colleagues at Swansea University have since found that

A 2016 survey of 2,000 people from AmeriSleep found that the most common nightmares involve falling, being chased, and death, despite the fact that Americans' most common fears revolve around the government, the environment, and money.

the emotional strength of the experiences we have when we are awake is linked to the content of our dreams, and the intensity of our dreaming brainwaves. Blagrove's team found that the intensity of a person's theta brainwaves was positively correlated with the number of diary items that appeared in their dreams. They also found that events that had a higher emotional impact were more likely to become incorporated into a person's dreams than blander, more neutral experiences. They concluded that the most intense dreaming activity occurs when a person's brain is working hard to process recent, emotionally-powerful experiences. This suggests that dreaming acts like overnight therapy to soothe the emotional impact of a person's experiences.

The next question is: what are dreams for? Freud in 1900 thought that dreams were simply to do with wish fulfilment. He thought they were manifestations of a person's deepest desires and anxieties. And, like everything else, he thought every dream topic represented the release of sexual tension. That's not the current theory. Deep non-REM sleep is associated with static thoughtful dreams, which are primarily driven by the hippocampus in the process of long-term memory consolidation and predominantly include memories of events 'as they happened' without the random novel combination of objects seen in REM sleep dreams.

Dreams that occur during periods of rapid eye movement (REM) sleep can be vivid and bizarre, and are the ones that people usually remember if woken up. REM sleep is also known as paradoxical sleep because of physiological similarities to being awake. REM sleep helps preserve certain types of memories, ie procedural memory, spatial memory, and emotional memory. And lack of REM sleep can inhibit learning. However, it has been suggested that acute REM sleep deprivation can improve certain types of depression. REM sleep occurs most just after birth, and decreases with age. According to Markov et al (2012), 80 percent of dreams occur during REM. And it's estimated that 95 percent of dreams are forgotten by the time a person gets out of bed. REM atonia is where there is almost complete paralysis of the muscles of the body. Obviously, the heart and lungs continue to work, but not other muscles. Interestingly, dreaming has also been shown to promote creative thinking (Llewellyn et al 2017).

Nightmares occur while a person is dreaming. The signs and symptoms of nightmares, according to the DSM-5, are:

- Repeated occurrences of extended, extremely dysphoric (a profound state of unease or dissatisfaction), and well-remembered dreams that usually involve efforts to avoid threats to survival, security, or physical integrity and that generally occur during the second half of the major sleep episode.

- On awakening from the dysphoric dreams, the individual rapidly becomes oriented and alert.

- The sleep disturbance causes clinically significant distress or impairment in social, occupational, or other important areas of functioning.

- The nightmare symptoms are not attributable to the physiological effects of a substance (eg a drug of abuse or a medication).

- Coexisting mental and medical disorders do not adequately explain the predominant complaint of dysphoric dreams.

Recurring nightmares may not be identical each time a person has them, but they will have similar themes. A person who is stressed or anxious may very well have more frequent nightmares. Becoming worried about a particular event or situation – including only dreamed of situations – can result in a person needing to dream about it again and again until the situation is resolved or loses its emotional content.

So, how can hypnotherapy help? The obvious way is to help a person deal with the stress or anxiety in their life – what we call stress bucket emptying. Hypnotherapy can also help a person to get a good night's sleep because sleep also helps to reduce stress levels. And lack of sleep may result from a person not wanting to sleep because they have nightmares. Hypnotherapy can also help a person with their negative thought patterns. Maybe, your accepted ideas of why a person (your boss, your staff, your children, your partner, etc) is behaving in a negative way towards you might be wrong. There might well be an alternative explanation for their behaviour that doesn't involve you.

A hypnotherapist might advise a person to not eat immediately before going to bed as a way of stopping nightmares. The therapist might suggest that people write down any concerns or worries before getting into bed, so they don't keep thinking about them while trying to go to sleep.

The therapist might discuss the recurring nightmare with a client and help them to rewrite the ending to a more positive one. Perhaps someone turns on the light and the bad guys run away. Perhaps the aliens are called away to invade a different planet. Perhaps the water stops filling up and escapes through a hole leaving you warm and dry. You get the idea!

The therapist will definitely show the client how to breathe in a way that will help a person to relax, eg 7-11 breathing or circular breathing. And the hypnotherapist will share recordings that will help the client to relax as they go to sleep or if they wake during the night.

These techniques can be successfully used with children and adults.

References:

https://psychcentral.com/news/2019/12/06/how-nightmares-help-us-face-our-fears-when-were-awake/152185.html

Markov D, Goldman M, Doghramji K (2012). "Normal Sleep and Circadian Rhythms: Neurobiological Mechanisms Underlying Sleep and Wakefulness". Sleep Medicine Clinics. 7: 417–426

Llewellyn S, Desseilles M. Editorial: Do both psychopathology and creativity result from a labile wake-sleep-dream cycle?. Front Psychol. 2017;8:1824. doi:10.3389/fpsyg.2017.01824

https://www.psychologytoday.com/gb/conditions/nightmares

https://elemental.medium.com/nightmares-are-good-for-you-according-to-scientists-e22fe3660166

https://www.newscientist.com/article/mg23931873-300-weve-started-to-uncover-the-true-purpose-of-dreams/

Are you lonesome tonight?

A look at loneliness and what to do about it.

As a solution-focused hypnotherapist and psychotherapist, I've been hearing more-and-more people mention that they have been feeling lonely lately. If that applies to you, you are not alone.

It's clear that a more-than-usual number of people have been feeling very lonely during their period in lockdown. For some people, like many elderly people, it was because they were on their own for many weeks. For others, there may have been people around them, but they weren't the 'right' people or doing the 'right' things. Some people may miss one particular person, such as a spouse, sibling, or best friend. And others may simply wish to be part of a wider social network that they could interact with.

But how common is loneliness generally? A 2018 report from the Office for National Statistics found that:

- In 2016 to 2017, there were 5% of adults in England who reported feeling lonely 'often' or 'always'.
- Younger adults aged 16 to 24 years reported feeling lonely more often than those in older age groups.
- Women reported feeling lonely more often than men.
- Those single or widowed were at particular risk of experiencing loneliness more often.
- People in poor health or who have conditions they describe as 'limiting' were also at particular risk of feeling lonely more often.

It works out that there are more than 900,000 people aged 65 and over in the UK reporting feeling lonely all, or most of the time.

According to de Jong-Gierveld and Raadschelders (1982), Duck (1992), and others, there are two levels of loneliness: chronic and transient. For people who are chronically lonely, their experience of loneliness is persistent, often extending to many years, and doesn't change with what the person is doing. It looks like the cause of the feelings is internal. They may feel the intensity of the loneliness vary over time, but it is always there. Transient loneliness, as its name suggests, is experienced for short periods of time, and is usually the result of a specific situation.

Courtney and Meyer (2020) published in the *Journal of Neuroscience* their findings that loneliness alters how the brain represents relationships. They suggested that the medial prefrontal cortex (mPFC) (near the very front of the skull) maintains a structured map of a person's social circles, based on closeness. People that struggle with loneliness often perceive a gap between themselves and others. This gap is reflected by the activity patterns of the mPFC. Courtney and Meyer used functional magnetic resonance imaging to examine people's brain activity while they thought about themself, close friends, acquaintances, and celebrities. Thinking about someone from each category corresponded to a different activity pattern in the mPFC. There was a different pattern for the self, for the social network (both friends and acquaintances), and for celebrities. The closer the relationship, the more the pattern resembled the pattern seen when a person is thinking about themself. These brain patterns differed for lonelier individuals. Activity related to thinking about the self was more different from activity related to thinking about others, while the activity when thinking about others was more similar across social categories. In other words, lonelier people have a 'lonelier' neural representation of their relationships.

Loneliness is definitely not good for you. For example, social isolation and loneliness:

- Have a negative effect on the activities of daily living of older peoples, including a person's ability to perform the normal daily activities that are required to meet their basic needs, fulfil their usual roles, and maintain their health and wellbeing.
- Impact on health-related physiology, eg blood pressure and reduced immune functioning.
- Lead to poorer sleep quality.
- Are associated with a greater risk of a person being physically inactive and smoking (both health-risk behaviours).
- Lead to lower self-esteem and limited use of active coping mechanisms.

Loneliness and social isolation affect physical health. For example, loneliness:

- Increases the likelihood of mortality by 26 percent.
- Has the same effect on mortality as the impact of obesity, and cigarette smoking and substance-dependency.
- Increases the risk of developing coronary heart disease and stroke.
- Increases the risk of high blood pressure.
- Is a risk factors for the progression of frailty.

In terms of mental health, loneliness:

- Puts individuals at greater risk of cognitive decline and dementia.
- Increases an individual's risk of depression.
- Is predictive of suicide (in older age) or self-harm.
- Is associated with poorer cognitive function among older adults.

Studies have shown that adolescents who are lonely may be more likely to use drugs or alcohol and become sexually active at an earlier age than their peers. Lonely teenagers are also more likely to engage in risky and unsafe sex or exhibit aggressive behaviour.

Things that can lead to feelings of loneliness include:

- The loss of a loved one
- A sudden breakup
- Single parenthood

- Retirement
- Moving to a new area or going away for college
- Health problems that limit a person's ability to socialize
- Surviving abuse.

The distress associated with loneliness can be significant and may lead to feelings of helplessness and hopelessness. People who are shy, experience social anxiety or are reluctant to take social risks, and they may be more likely to describe themselves as lonely and may have difficulty forming lasting and satisfying relationships.

It should be remembered, that while many people who feel lonely are physically alone, not everyone who is alone feels lonely. Some people simply choose to have few social connections. If a person chooses to be alone, they may well enjoy and welcome the solitude.

One of the issues with loneliness is that many people are reluctant to admit it – they feel that it is a sign of weakness. The obvious solution to feelings of loneliness is go out and meet people, but that can be hard. Obviously, during lockdown that was very difficult!

It's worth noting that being alone isn't all bad. Julie Bowker, assistant professor in the Department of Psychology at the University of Buffalo in 2017 looked at 295 people who were unsociable due to non-fearful preferences for solitude. They weren't shy or avoiding other people. The study found that being alone can improve a person's state of mind, increase their creativity, and decrease stress and depression.

So, what can people do to stop feeling lonely? The charity Mind offers some suggestions:

- Take it slowly – go somewhere where there are other people, but you're not expected to interact.
- Make new connections – join a class or volunteer.
- Try peer support – try a befriender service or join an online community, eg Elefriends (https://www.elefriends.org.uk/).
- Try to open up – reach out to someone or share a post on social media.
- Try talking therapies – like hypnotherapy or CBT.
- Social care – The Care Act 2014 places general obligations on local authorities to promote wellbeing and to prevent social care needs from arising.
- Be careful when comparing yourself to others – people only post on social media and tell you about the high points in their life. It may not always be like that.
- Look after yourself – get enough sleep, eat healthily, get some exercise, get outside, spend time with animals. And avoid drugs or alcohol.

When you visit a solution-focused hypnotherapist, we can help with any issues associated with loneliness such as depression, anxiety, or anger issues. We can help people to sleep better. We can help you to relax. And we can suggest that you treat any attempts to socialize as experiments. If an attempt at socializing, eg joining a badminton club, doesn't work, then the information you get from the result of that experiment can simply help you when you plan your next experiment, eg joining a bridge club (or whatever). The result doesn't reflect in any way on you personally. And the very fact that you are able to share your inner thoughts with your hypnotherapist, may very well make it easier for you to talk to any new people you meet.

Loneliness is a big problem, and it, in many ways, is a hidden problem. The good news is that solution-focused hypnotherapy can help.

References:

https://www.ons.gov.uk/peoplepopulationandcommunity/wellbeing/articles/l onelinesswhatcharacteristicsandcircumstancesareassociatedwithfeelinglonely/ 2018-04-10

https://www.psychologytoday.com/gb/blog/web-loneliness/201404/treating-loneliness-its-more-just-meeting-others

https://www.eurekalert.org/pub_releases/2020-06/sfn-lay060920.php

https://www.campaigntoendloneliness.org/threat-to-health/

https://www.goodtherapy.org/blog/psychpedia/loneliness

https://www.mind.org.uk/information-support/tips-for-everyday-living/loneliness/tips-to-manage-loneliness/lonelinesswhatcharacteristicsandcircumstancesareassociated withfeelinglonely/2018-04-10

https://www.psychologytoday.com/gb/blog/web-loneliness/201404/treating-loneliness-its-more-just-meeting-others

https://www.eurekalert.org/pub_releases/2020-06/sfn-lay060920.php

https://www.campaigntoendloneliness.org/threat-to-health/

https://www.goodtherapy.org/blog/psychpedia/loneliness

https://www.mind.org.uk/information-support/tips-for-everyday-living/loneliness/tips-to-manage-loneliness/

http://www.buffalo.edu/ubnow/stories/2017/11/bowker-social-withdrawal.html

Job seekers and hypnotherapy

How to help unemployed people looking for work.

One expected consequence of the pandemic is that a large number of people will become unemployed and will be looking for work. It's likely that there will be fewer jobs available for them as the economy shrinks. How can solution-focused hypnotherapists help these people?

Many people will be worried because of the loss of a regular income. Some will feel that they have lost their identity. And some will be feeling quite depressed. They may dread the whole process of looking for potential jobs, applying for them, and then never hearing back or being rejected. Job hunting can be challenging and exasperating. People may well feel that they have lost control over their life. And not having a job may make them feel ashamed or embarrassed. They may be worried about paying bills and everyday expenses. And constant rejection may make them feel a failure.

As well as depression, their self-esteem and feelings of self-worth may be rock bottom. And they may be feeling anxious all the time. They may find themselves brooding over what's gone wrong in the past or how awful things will be in the future. And they may be comparing themselves with friends, family, and other ex-employees from their old company, who all seem to be doing so much better than they are at finding work. This may well be leading them to lying awake for long periods in the night worrying. It may also be leading them to put on weight and they may be drinking more than usual.

So, what can their friendly neighbourhood solution-focused hypnotherapist do to help? Anything emotionally bad that happens to a person goes into their metaphorical stress bucket. The first thing to do is to help the client to empty their stress bucket. This can be done by helping them to relax and to see the bigger picture. To help them get out of their primitive brain and move back into their intellectual brain where they are better able to come up with innovative ideas. Maybe they should start that baking business they always dreamed of (for example). Or maybe they shouldn't. It will be a logical decision whatever they choose.

Solution-focused hypnotherapists can help them with sleep issues. We can give them plenty of advice about what things to do before bed that will help them to sleep, eg warm bath, regular bedtime, make sure the room is dark, (see Insomnia page 11), etc. And we can advise them on things not to do, like looking at screens, exercising just before bedtime, drinking alcohol or coffee. We can recommend that they get up at the same time every day. And we can recommend that they keep to a regular routine.

We can recommend that they focus on the 3Ps – positive thoughts, positive actions, and positive interactions. We do this by asking what's been good every time we see them, what were their sparkling moments. Too often, people begin to self-isolate when they feel low or depressed. We can encourage them to continue to see other people and talk to them (even on Zoom). It will be good for our client to do that. Maybe, they will have to listen to the success stories of others and feel envious of their success, but they can also let everyone know that they are still looking for work and hopefully use that network of people to put them in touch with someone who is looking for staff. The fourth P is purpose. Your client may feel that losing their job means that their life has no purpose. In fact, getting a new job is their new purpose.

What can be done to raise a person's self-esteem – their opinion of themselves? The first thing to do is find what negative beliefs the person has of themselves and challenge them. Is it always true that they are xxx? Ask them about the things they are good at. Get the client to write down these good things and help them to think of more. If people say nice things about them, get them to write down those as well. Also, encourage them to recognize the things they have already achieved. Encourage clients to take part in activities that they enjoy. Encourage them to remember times when they have been happy, relaxed, and at peace. And encourage them to spend time with positive, supportive people rather than negative people. And, also, encourage them stop comparing themself to other people, particularly on social media.

Praising people and offering positive approval works well on people with high self-esteem, but for people with low self-esteem, it makes them feel worse. Sometimes, you will find that they have all-or-nothing thinking – everything is black and white. They may feel that they don't deserve a job. You need to work with them challenging their beliefs and help to move them forward. They may feel that they are not as good as someone else. In that case help them to see their positive differences, and explain that how they are different is a good thing. Ask them to look out for triggers that trigger feelings of stress or low self-esteem. Once these are known, they can be avoided, or you can help your client to interpret them differently.

Encouraging unemployed clients to volunteer will not only increase their own self-esteem, but will look good on their CV to show that they are still used to some kind of work environment.

You can discuss emotional resilience with them and what techniques they have used in the past to feel more self-confident. Or ask them when don't they feel anxious, stressed, or have low self-esteem?

You might encourage them to take some exercise each day. Exercise, obviously, is good for the body and it encourages the growth of new brain cells. You might also look at what they eat, when they eat, and why they eat. Encouraging them to keep a food diary and perhaps putting a mirror on the biscuit cupboard could be the first step towards helping them to get in control of their weight. Similarly, they might keep an alcohol diary so they can see, during the day, how much they are drinking and when. More work can be done to help them get back in control of their drinking.

What about with the rest of their life, what suggestions could you make? As I said, I would treat finding a job as a job itself. That means getting up early and getting dressed. There is a saying, "No matter how you feel, get up, dress up, show up, and never give up". It applies in this situation. And you might advise them to give their day structure, and let job hunting end at 5:30pm. Spending all day looking for work can lead to burnout. They could use part of their working day to learn a new skill or extend their existing skill set. They could attend local networking events (which may well be online) during their working day and tell people that they are looking for their next opportunity. Part of the day can be spent reviewing and updating their CV. Perhaps it will need tweaking for some potential jobs to focus more on the skills that job needs. Perhaps time can be spent tailoring application emails for specific jobs. Or time can be spent preparing emails for companies that aren't currently advertising vacancies, but might be a good fit for your client's skill set.

It's important that they look after themselves. That's why a good night's sleep, exercise, and a healthy diet are so important. It's also important to celebrate their successes, no matter how small, on the road to employment. And talk about not having a job with the word 'yet', ie "although you don't have a job yet, xxx". That way, you're framing it as something that's temporary not permanent. You're helping your client to see that there is light at the end of the tunnel.

And you should encourage them to seek help. It's often better if a second pair of eyes looks at a CV or covering letter just to spot any typos and make sure that the ideas expressed make sense to a person who doesn't know your client.

And when their 'working' day is over, they should be encouraged to do those activities that they enjoy and make them feel good. This will put them in a better frame of mind for tomorrow. Plus, continuing with hobbies may give your client something to talk about at the interview.

If your client follows your advice, they will no longer be feeling depressed, their stress bucket will be emptying each night by itself. They will be able to celebrate others' successes and not feel bad about it. They will have stopped brooding, and their self-esteem will be much higher. They will be sleeping well, exercising enough, and enjoying a healthy diet. They will be thinking positively, acting positively, and interacting positively with people. They will give the best impression they can to potential employers. And if they don't get the jobs they apply for, it will be the employers' loss.

The resilient client

Understanding how to help clients be more resilient.

Resilience, in the psychological sense, is a person's ability to mentally or emotionally cope with a crisis or to return to pre-crisis status quickly. You can think of it as the ability to remain calm during crises/chaos and to move on from the incident without any long-term negative consequences. Resilience doesn't mean that a person isn't affected by what's going on around them, it's just that they have developed psychological and behavioural capabilities to cope. People who have suffered major adversity or trauma in their lives commonly experience emotional pain and stress.

It used to be thought that some people had the characteristic or trait of resiliency and they were therefore better able to cope with difficult situations than people who didn't have that particular characteristic. Nowadays, there's a different idea. Resilience is now thought of as a process. So, you're faced with an adverse condition, how do you respond? It seems people generally react in one of three ways:

- Erupt with anger
- Implode with overwhelming negative emotions, go numb, and become unable to react
- Simply become upset about the disruptive change.

It seems that resilient people choose the third option. The first two options lead to people rejecting potential coping methods, blaming others, and seeing themselves as victims. Negative emotions such as fear, anger, anxiety, distress, helplessness, and hopelessness decrease a person's ability to solve the problems they face and reduces their resiliency. A resilient person will adapt to the adversity and be able to cope.

A lot of the research into resilience has been around stress and how people adapt to it. It seems that repeated exposure to stressors in daily life can promote more resilience. The more challenges we face, the better our resilience strategies become. It also seems that developing resilience by adapting is positively associated with increased happiness. Hope, optimism, and self-efficacy seem to be the kinds of characteristic that help a person to adapt to challenges. They also enhance a person's ability to cope and thrive after dealing with an event or challenge.

Strangely, failure can be a good thing in developing resilience. Young scientists who experienced a significant setback early in their career actually went on to greater success than scientists who had seen early successes! Accepting mistakes builds better emotional regulation. And, analysing and accepting a setback can provide lessons that can prevent future failures from being repeated.

So, what characteristics would a resilient client have? Research suggests they would have:

- The ability to make realistic plans and be able to take the steps necessary to follow through with them
- Confidence in their strengths and abilities, and a positive self-image

- Communication and problem-solving skills
- The ability to manage strong impulses and feelings.

There also seems to be a strong link between positive emotions and resilience. Research has found that maintaining positive emotions while facing adversity promotes flexibility in thinking and problem solving. Positive emotions help a person to recover from stressful experiences and encounters.

People who tend to approach problems with cognitive reappraisal, humour, optimism, and goal-directed problem-focused coping seem to strengthen their resistance to stress.

A study found six main predictors of resilience among high achieving professionals. These were: positive and proactive personality, experience and learning, sense of control, flexibility and adaptability, balance and perspective, and perceived social support. These high achievers also engaged in many activities unrelated to their work such as hobbies, exercising, and organizing meet-ups with friends and loved ones.

Lots of studies show that the primary factor for the development of resilience is social support, ie how much access they have to, and use they make of, strong ties to other individuals who are similar to them. Other factors associated with resilience include: the capacity to make realistic plans, having self-confidence and a positive self-image, developing communications skills, and the capacity to manage strong feelings and impulses. Forgiveness also seems to increase resilience.

The American Psychological Association suggests "10 Ways to Build Resilience", which are to:

1 Maintain good relationships with close family members, friends, and others.

2 Avoid seeing crises or stressful events as unbearable problems.

3 Accept circumstances that cannot be changed.

4 Develop realistic goals and move towards them.

5 Take decisive actions in adverse situations.

6 Look for opportunities of self-discovery after a struggle with loss.

7 Develop self-confidence.

8 Keep a long-term perspective and consider the stressful event in a broader context.

9 Maintain a hopeful outlook, expecting good things and visualizing what is wished.

10 Take care of one's mind and body, exercising regularly, and paying attention to one's own needs and feelings.

Cavenaugh et al (2000) looked at the 'challenge-hindrance stressor framework', where people who view problems with curiosity are more likely to solve the issue and move forward, rather than be defeated by the issue itself. They view problems as an opportunity for growth and as a chance to improve themselves. People with a hindrance perspective see challenges as something that happens 'to them', whereas people with a challenge perspective see problem as something that happens 'for them'.

So, how can we help clients? Firstly, we can help them create and maintain healthy habits such as getting enough sleep, eating a healthy diet, and exercising (or at least regular movement). These can reduce the physical effects of stress on the body and emotional distress in the brain. And that boosts a person's resilience – both when faced with an immediate challenge or threat, and when coping with future difficulties.

A 2009 study by Phillippa Lally *et al* published in the *European Journal of Social Psychology* found that it takes from 18 to 254 days to create a new habit, depending on the habit.

Techniques such as the Miracle Question, trance, and use of metaphor, can help clients to visualize modifying less beneficial behaviours, eg drinking too much, which will reduce stress and increase emotional and physical resilience. Trance will also help them to relax and get back into their intellectual brain.

Interacting with other people can help build resilience. In fact, statistically, resilience is highly correlated with peer support and group cohesion. It seems that groups with high cohesion tend to experience a lower rate of psychological breakdowns than groups with low cohesion and morale. We can encourage clients to spend time with people who have similar values, interests, and shared goals.

And we can help clients to cultivate a positive, optimistic, curious, and solution-oriented mindset.

References:

https://en.wikipedia.org/wiki/Psychological_resilience

https://positivepsychology.com/resilience-in-positive-psychology/

https://www.psychologytoday.com/us/basics/resilience

https://psychcentral.com/lib/what-is-resilience/

https://www.hcplive.com/view/10-ways-to-build-resilience

https://onlinelibrary.wiley.com/doi/abs/10.1002/ejsp.674

Online therapy – as good as the real thing?

A look at online therapy and its benefits

Since the Coronavirus pandemic started, hypnotherapists have moved to online working only – this means using things like Zoom or WhatsApp for video calls or simply talking on the phone. The worry that many potential customers (for this kind of therapy) have is whether online hypnotherapy works as well as actually sitting in the same room as your therapist. Will it work if you can't look into the eyes of the hypnotist? Will you really be able to stop smoking – or whatever change they envisage – if all you're doing is sitting in your dining room looking at a computer screen or your phone for a couple of hours?

The good news is that the answer is 'yes'.

On the downside, you will need to have a reliable Internet connection. And you will need to have somewhere quiet where you can speak freely and relax without interruption.

On the plus side, you save time by not needing to travel to see the therapist and get home afterwards. You also save on the cost of travel – whether that's petrol or bus fares. And you don't need to find somewhere to park and pay for parking. If you have mobility issues, this is a big plus. Of course, if you live in a remote area, then online therapy at any time makes life so much easier. Also, if you are worried about going outside – whether that's because you want to continue self-isolating for a while or because you have agoraphobia – your worries disappear. It also works well for people who find it difficult to accommodate visiting a therapist into their busy lives, such as key workers on shift, parents, and full-time carers. And there's no chance of getting someone else's germs. So, it's ideal for people who feel more safe-and-secure being at home rather than going out to a clinic.

Another big plus is that you can work with a therapist anywhere in the country. You're not restricted to local therapists. So, if you wanted a solution-focused hypnotherapist because you like the idea of working that way, you can choose anyone who is qualified and on the AfSFH register (https://afsfh.com/find-a-therapist) – no matter where they are based. Certainly, it is always worth choosing a therapist who belongs to an accredited organization, like the Association for Solution Focused Hypnotherapy, and who is also a member of the Complementary and Natural Healthcare Council (CNHC). The CNHC is the UK regulatory body that provides a voluntary register of complementary, rather than alternative medicine, therapists.

These days, people shop online – whether that's Amazon, their local supermarket, and much else. They play online games. They 'google' plumbers and gardeners, etc. They book holidays online. So much of life is online that seeing a hypnotherapist is not that much different.

Certainly, any hypnotherapist will tell you that the number of people asking about online hypnotherapy is growing.

Enquiries for online hypnotherapy sessions are growing in popularity. And online hypnotherapy can be very easy to access, even for people who previously might have described themselves as not very IT savvy. The technology, using Zoom and similar products, makes it all very straightforward and nothing to worry about. Most of the online meeting technologies are encrypted, so the communication and the whole session remain private and confidential.

You also need to ensure that the technology works at your end, ie there is a high-speed broadband link, and the camera and speakers on your laptop or phone will work in a therapy situation. Your therapist will probably test this before the first session. In the event of something going wrong, eg a power cut, the phone line being disrupted, or anything else, it's a good idea to have a phone near you that the therapist can call. But if you don't have these things, then you can simply talk on the phone. For online/phone sessions, payment must usually be made before each session starts. You will be given bank details in plenty of time to transfer the payment.

Since the lockdown started, many people have enjoyed online hypnotherapy, and there is plenty of anecdotal evidence of how well it works, but some people are still looking for evidence that an online therapy session is as good as a face-to-face session. The good news is that there is already some clinical evidence showing the efficacy of online hypnotherapy. For example, there's a 2014 study entitled "Internet-based versus face-to-face cognitive-behavioral intervention for depression: A randomized controlled non-inferiority trial" (https://www.sciencedirect.com/science/article/abs/pii/S0165032713005120) and published in the *Journal of Affective Disorders*. It found that treating depression using an Internet-based intervention is equally beneficial as regular face-to-face therapy. The study also reported: "However, more long-term efficacy, indicated by continued symptom reduction three months after treatment, could only be found for the online group." Similarly, a 2018 study entitled, "SKYPE HYPNOTHERAPY FOR IRRITABLE BOWEL SYNDROME: Effectiveness and Comparison with Face-to-Face Treatment" (https://www.tandfonline.com/doi/full/10.1080/00207144.2019.155376 6) and published in the *Journal of Clinical and Experimental Hypnosis* said: "This study shows that Skype hypnotherapy is highly effective in refractory IBS".

So, the anecdotal and the experimental evidence go to show that online hypnotherapy is definitely as good as the face-to-face version, and may, in some cases, be better! If you had concerns about giving it try, join the hundreds of people who are already benefitting from online hypnotherapy and let it help you.

Staying safe online

Some tips for hypnotherapists to avoid being hacked.

If you're working from home or in a small local clinic environment, you feel safe. Unfortunately, hackers are trying to get your information and your clients' information all the time. And losing personally-identifiable information for customers could lead to massive fines. In addition, no-one wants to pay a ransom to get back their data.

So, what can you do? Firstly, always make sure that the software you use on your laptop and phone is up-to-date. Hackers are always finding security issues with software so they can access the data on the device. And vendors are regularly patching the software to prevent a breach occurring. If you haven't already, install, use, and keep updated antivirus, antispyware, and anti-ransomware software. Use the firewall on your laptop too.

Never respond to phishing attacks. These are emails, text messages, etc that get you to click on links and fill in your details on a spoof (hacker run) website. The hackers will take the details you supplied and use it on real websites, eg your bank's details.

Don't overshare on social media. No-one wants to come home to find their house burgled because they told everyone on Facebook that their whole family was away for two weeks. And if you do use your children's or your dog's name as your password, don't publish that information on Facebook for all to see. In fact, make your Facebook account secure so no-one can see the year you were born (you still want them to wish you a happy birthday!), and you only want friends to see your photos. Remember, posting lots of information on social media means that hackers know your location, your date of birth, your interests, and much more. It makes a phishing attack far more likely, as well as giving them lots of information when trying to guess the answers to your security questions. It can lead to identity theft, where they pretend to be you and run up bills that you may become liable for.

When you sign up to a new website, don't use the same password as for every other website. So many large sites – like LinkedIn – have been hacked and their data stolen and made available on the dark web. So, if hackers know what your password is for one site, they can guess you're using the same one for your bank!

Similarly, use a different answer to the "what's your mother's maiden name" question for each website. It's probably best to keep a record of all these passwords and authentication recovery questions. And password protect that file.

If you ever buy anything online, make sure that the web address (URL) of the site begins 'https'. And, even then, double-check that it really is a real shopping site.

Going back to emails, watch out for emails that seem to come from your bank or Microsoft or any other organization that you do legitimately have a relationship with. Check the 'reply to' address. Is it ever-so-slightly misspelled? Then it's not genuine. Or does it include the right company name, but not as part of the domain? That's another hacker ploy. Or is it the real brand name, but with some extra letters at the end? You guessed it, it's spoof. Other tricks include using strangely encoded domain names, short URLs, incredibly long URLs, and URL redirects at the end of what looks like a genuine address, or just a completely different URL!

Never click on an attachment in an email that feels in any way suspicious to you. These often contain malware – software that gives the hackers access to your computer. They could install key loggers to record your every keystroke as, for example, you log into your bank account. It could be ransomware and you'll have to pay, usually in Bitcoins, to get access to your data, And, once you've paid, they still may not give you access. That does show the importance of backing up your data regularly and frequently. Hackers may exfiltrate (that's the technical word for 'take a copy of') your data, which they can sell on the dark web and which may well lead to you paying a large fine. Note: a text file (.txt) is safe to click on, but anything else could contain a payload that you won't want on your computer.

Don't click on links embedded in an email without first hovering your mouse over the link. This will display the actual URL, ie where the link will take you. It may very well not be where you expected. Obviously, if the address is OK, then you can click on the link.

Don't respond to spam in any way. Don't even click on the unsubscribe button. All that does is prove to the spammers that yours is a real address – and they will then start their phishing attacks. Simply delete the email.

Other things that should alert you that an email may not be genuine are:

- You don't recognize the sender's email address as someone you usually communicate with.
- The sender's email address is from a suspicious domain (like asfsh.support.com)?
- It's an unexpected or unusual email with an embedded hyperlink or an attachment from someone you haven't communicated with recently.
- You've been cc'd on an email sent to other people, but you don't know them.
- You receive an email with a subject line that is irrelevant or does not match the message content?
- The email header says something like: Covid-19 cure; HMRC repayment; Black Lives Matter; New client; or anything else in the news or likely to attract your interest.
- An email appears to be a reply to something you never sent or requested?
- An email asks you to click on a link or open an attachment to avoid a negative consequence or to gain something of value.
- An email is out of the ordinary, or contains bad grammar or spelling errors
- An email invites you to look at a compromising or embarrassing picture of yourself or someone you know.

In your home, make sure that IoT (Internet of Things) devices, like smart speakers, appliances, etc, use strong unique passwords whenever possible.

If you do go out, for example at your favourite coffee shop, don't use public Wi-Fi when accessing confidential information, eg your bank. And don't allow your devices to auto-join unfamiliar Wi-Fi networks.

Always disable automatic bluetooth pairing. Bluesnarfing and Bluejacking are the hacking terms for using someone else's bluetooth connection without their knowledge. Bluejacking usually involves sending a message to a target device, Bluesnarfing is where information is stolen from the target device.

Never use unknown USB devices such as a memory stick. Who knows what malware might be on it! Always lock your laptop if you leave for a moment, eg to go and buy another coffee or muffin!

And if you do all these things, you and your hypnotherapy business should stay safe.

The opposite of stress

A look at the effects of kindness.

David Hamilton, author of *The Five Side Effects of Kindness: This Book Will Make You Feel Better, Be Happier & Live Longer*, reckons that kindness is the opposite of stress. And tells us some of the benefits of being kind. What he suggests is that all the negative effects of stress on the body can be reversed by acts of kindness.

Hamilton suggests that a side-effect of being kind is that it makes you happier (stress makes you unhappy). Jonathan Haidt calls the feeling 'elevation' – the positive feelings we get from being kind. Allan Luks, author of *The Healing Power of Doing Good* called the feeling "helper's high" – the warm feeling you get from being kind.

Studies have found that people who volunteer for charities and for other groups and organizations, are more resistant to depression, ie the likelihood of them becoming depressed is less than average. It's as if kindness increases a person's resilience.

In one study, researchers surveyed people's stress levels over a few weeks and counted the numbers of acts of kindness per day they performed. The results showed that on days when people had been kinder, their stress levels were lower, and on days when their levels of stress were higher, they performed fewer acts of kindness. The conclusion was that being kind and the feelings that you get, the "elevation", the "helper's high", somehow form a buffer, making the stressful events more manageable. And that's one of the reasons why the likelihood of becoming depressed and anxious is lower in people who consistently perform acts of kindness.

One way to test this idea is to ask people to either behave as normal or to perform acts of kindness. The result from many different experiments like this is that people who are being kind on purpose tend to be happier at the end of the study than those who are behaving as normal. So, kindness actually makes us feel happier.

The effects of kindness and compassion go beyond just a mental and emotional sense of resilience, they also have a positive effect on the nervous system and the cardiovascular system.

Compassion is the feeling that usually motivates a kind act, and it has anti-inflammatory properties. Compassion stimulates the vagus nerve, which controls the inflammatory reflex, which helps reduce inflammation.

When a person feels stressed, their brain and body are full of adrenalin and cortisol – ie stress hormones. When a person is being kind, they feel the effects of oxytocin (the love hormone or cuddle chemical).

Oxytocin is associated with feelings of warmth and connection, or having any kind of strong, warm emotional contact with a person. One of the things that oxytocin does is act on receptors on the lining of arteries. It then generates nitric oxide, which softens the walls of the arteries. And, as the arteries soften, they widen. And as they widen, the heart doesn't have to push quite so hard to get the blood through. So, a person's blood pressure drops. Because of this, oxytocin is called 'cardioprotective'. So, being kind is good for your heart.

Other studies have shown that levels of coronary artery calcification depend on how much kindness, warmth, compassion, love, and affection a person shows. The more acts of kindness, the lower their levels of calcification. So, being kind protects your coronary artery.

Studies have also shown that higher levels of oxytocin can reduce levels of inflammation and oxidative stress. (Oxidative stress is an imbalance of free radicals and antioxidants in the body that can lead to cell and tissue damage.) These are two processes that are linked with cardiovascular disease.

Kindness also has a positive effect on the immune system. This can be measured using markers such as secretory immunoglobulin A (s-IgA). In one experiment, volunteers watched a video of Mother Theresa on the streets of Calcutta, carrying out acts of kindness and compassion. Simply watching acts of kindness and compassion increased their levels of s-IgA. The increase was attributed to how watching the video made them feel. Even talking about the video later kept their s-IgA levels higher because of how it made them feel.

Another study got people to contemplate kindness and compassion for 5 minutes. That also resulted in increased levels of s-IgA.

Kindness also slows ageing. Stress can speed up ageing, and lots of stress can cause visible signs of ageing, like increased wrinkling, because it creates oxidative stress, which produces free radicals (which antioxidants in our diet combat).

Studies of skin cells show that when levels of oxytocin are high, the levels of oxidative stress are lower. So, there are fewer signs of ageing.

One other thing to note, you can't get a positive side effect of kindness unless you mean it. Your body won't produce oxytocin and there will be no physical benefit.

Lastly, a mention of self-compassion, that feeling of understanding your own humanness. Self-compassion is like giving yourself a hug.

I'm sure it's worth reminding clients to be kind as a way of overcoming the stress that they may well be feeling. So, what can we suggest that they do?

One of the 4Ps is positive interaction with other people. So we can advise people to connect with other people (and increase their oxytocin levels).

For homework, suggest that for the next seven days, clients make an effort to strike up conversations, to listen to people, to offer help and support where they can. They should smile at people and say hello. Or, they must be intentionally kind. They have to look for opportunities and act on them. For the seven days, they must do something different each day. They must do something that pushes them out of their comfort zone at least once during the week. And one act of kindness has to be completely anonymous. They can then try 21 days of kindness or 365 days!

Positive psychology offers many other ways of feeling kind, eg keeping a gratitude journal, writing a letter of thanks to someone, thinking of three good things that happened that day, or reminiscing positive memories.

Anything that's the opposite of stress has got to be good.

Bored to tears

A look at the positive and negative side of boredom.

It all started during lockdown with a client who said she was "so bored". I quickly suggested that we put that to one side for now and asked her to tell me what had been good. And the session proceeded in the usual way. But afterwards, I started to wonder about boredom. What is it for? Is it a good thing or a bad thing? How many people experience that feeling of being bored? And is it only boring people who feel bored? Let's see what the research shows us.

Wikipedia tells us that boredom is an emotional and occasionally psychological state experienced when an individual is left without anything in particular to do, is not interested in their surroundings, or feels that a day or period is dull or tedious.

The word 'boredom' was first used in print by Charles Dickens in *Bleak House* in 1852. Lady Deadlock is described as being "bored to death" by her marriage. 'To be a bore' meaning to be tiresome or dull has been around in print since 1768. The word 'bore' (noun) meaning a thing causing ennui or annoyance has been around since 1778. The French word for boredom, ennui, has been borrowed into English since the 1660s.

A survey of 2,000 Americans found that for the average adult, more than a third of their year is spent feeling bored. The definition of boring they used was that a day involved simply no fun at all. After averaging out responses from all the participants, they found that Americans experience a staggering 131 boring days each year. It's perhaps worth noting that the survey, conducted by OnePoll, was sponsored by Bowlero, a bowling alley chain.

What is boredom for? Evolution seldom leaves a species with a characteristic that is useless. The philosopher Andreas Elpidorou suggests that boredom is a "push to act". In the moment, we feel mentally unoccupied, and this state is deeply distressing. Our preferred state is to be optimally engaging with the world to showcase who we are and what we are capable of. Boredom acts as a call to action to achieve that end. People want to be the author of their own actions and choices. When that authorship, or agency, is removed from them, boredom ensues.

Heather Lench, from Texas A&M University, suggests that boredom lies behind curiosity. Boredom, she thinks, stops people doing the same old same-old and pushes them to try to seek new goals or explore new territories or ideas. However, she warns, searching for an escape can sometimes lead people to take risks that eventually harm them!

What kinds of people get bored? John Eastwood, from York University in Canada, found that people with two distinct personality types tend to suffer from ennui. He suggested that boredom often goes with a naturally impulsive mindset among people who are constantly looking for new experiences. For them, the world is chronically under-stimulating – so they feel bored. The second personality type belongs to people who find the world to be a fearful place, and so they shut themselves away and try not to step outside their comfort zone. This can lead to chronic boredom.

Just because something is objectively meaningful doesn't mean it feels that way to the people doing it all the time. Anaesthetists often describe their work as 99% boredom and 1% panic. Similarly, air-traffic controllers have been suspended for watching DVDs at work. It seems that meaningful work can be boring if the person performing it finds it too hard or too easy. Once that happens, individuals might have problems staying focused.

So, if you are feeling bored, what can you do about it? Well, you could:

- Read a book or do some mindful colouring in.
- Write a list – of places you want to visit or tasks that need to be done or your favourite Christmas songs. Or, you could write a list of all the things you are grateful for.
- Make something as a gift for someone.
- Exercise or do yoga or go for a walk in the countryside or dance round the room.
- Socialize – interacting positively with other people is always good.
- Learn a new skill – haven't you always wanted to play the saxophone or speak Mandarin?
- Do some gardening
- De-clutter your wardrobe – or your whole home.
- Watch old comedy programmes that make you laugh.
- Look through old photos and remind yourself of good times in the past.

If you are feeling bored, is that bad for you? Research has shown that being bored can be associated with a number of negative outcomes, including: poor mental health; increased rates of depression, anxiety, and aggression; and problems with alcohol and drug abuse. A 2010 study by Annie Britton and Martin J Shipley found that British civil servants who reported high levels of boredom in their job were more likely to die from heart disease decades later. A 2011 study by Ulrike E Nett, Thomas Goetz, and Nathan C Hall looking at boredom in schools found that boredom is associated with poor performance. For teens generally, boredom is a risk factor for substance abuse, risky sexual acts, and even vandalism.

Is being bored always bad for you? A 2019 study by Guihyun Park, Beng-Chong Lim, and Hui Si Oh looked at why being bored might not be a bad thing. They gave one group of people a boredom-inducing task – they had to methodically sort a bowl of beans by colour, one by one. A second group were given a more interesting craft activity. Next both groups were given an idea-generating task in which they had to come up with excuses for being late that wouldn't make someone look bad. What they found was that the group who had been bored outperformed the artists both in terms of idea quantity and quality. The ideas suggested were later ranked by objective outsiders who assigned uniqueness scores to each one.

Sandi Mann, a senior psychology lecturer at the University of Central Lancashire and author of *The Upside of Downtime: Why Boredom Is Good* also researched whether there were positive outcomes from being bored. She came up with five positives. They were:

- Boredom sparks creativity – when people are bored and looking for stimulation, if they can't find it, they make up their own. That's what daydreaming is.

- Boredom allows a person to engage in self-reflection – being bored gives a person time to sit and think about their current circumstances. This self-reflection often leads to real changes and improvements in a person's life.

- Boredom is good for your mental health – daydreaming can provide a brief escape from day-to-day life, So, people should step away from screens and work generally because they can strain mental health. This break can be a valuable opportunity to recharge.

- Boredom can make you a better person – boredom can make a person altruistic. It's thought that when a person is bored, they lack perceived meaning. So, people search elsewhere to re-establish the value of their life, which, in turn leads them to prosocial behaviours like giving to charity.

To become bored, the recommendation is to pick an activity that requires little or no concentration, eg walking a familiar route, swimming laps, or just sitting with your eyes closed, and simply letting your mind wander, without music or stimulation to guide it. No screens or devices should be nearby.

So, bored people are searching for something to stimulate them that can't be found in their immediate surroundings, and this can lead them to start daydreaming. This process allows different connections in the brain to be made. And this leads to people having new ideas.

When a person is consciously doing something, they're using the executive attention network, the parts of the brain that control and inhibit their attention. When their minds wander, they activate the default mode network. Default mode is often used to describe the brain at rest, ie when it's not focused on an external, goal-oriented task.

The areas of the brain that comprise the default mode network are the medial temporal lobe, the medial prefrontal cortex, and the posterior cingulate cortex. They are active when a person is thinking about others, thinking about themselves, remembering the past, and planning for the future.

Daydreaming is involved in a wide variety of skills, from creativity to projecting into the future. When the brain is in default mode, it's still using about 95 percent of the energy it uses when thinking. So, what can we conclude? Boredom is bad for you except when it's good for you! Being bored can lead a person to coming up with great ideas and for re-evaluating their life. It can indicate that they are recharging their batteries away from the stresses of work. However, when people are bored, they will look for things to reduce those bored feelings – and that can lead to drug abuse, substance abuse, smoking, and alcohol abuse, which can all have a negative effect on a person's life.

If you have a client who says they feel bored, help them to appreciate the benefits of how they feel and avoid the pitfalls.

References:

https://en.wikipedia.org/wiki/Boredom

https://www.studyfinds.org/no-fun-americans-mired-boredom-131-days-year/

https://theconversation.com/6-things-you-can-do-to-cope-with-boredom-at-a-time-of-social-distancing-134734

https://www.psychologytoday.com/gb/blog/the-engaged-mind/202006/why-does-boredom-exist

https://academic.oup.com/ije/article/39/2/370/684049

https://www.sciencedirect.com/science/article/abs/pii/S0361476X10000548

https://journals.aom.org/doi/10.5465/amd.2017.0033

https://time.com/5480002/benefits-of-boredom/

https://www.goodnet.org/articles/5-scientific-reasons-boredom-actually-good-for-you

https://www.psychologytoday.com/gb/blog/brainstorm/201709/unplug-get-bored-create

Diet and depression

A look at some of the evidence that changing your diet can reduce the symptoms of depression.

No-one really knows what causes depression or what prevents depression. Research by B L Beezhold *et al* and published in the *Nutrition Journal* in 2010 looked at Seventh Day Adventists. They were given a survey to assess their mood. There were 78 omnivores and 60 vegetarians in the study, which found that the vegetarians had a better mood.

A larger study of 620 people in 2015 by B L Beezhold *et al* and published in the *Nutritional Neuroscience* journal looked at stress and anxiety. The vegans reported less anxiety and stress than the lacto-ovo vegetarians (ie vegetarians who ate dairy products and eggs).

Another study by Sanchez-Villegas *et al* (2009) compared a Mediterranean diet with other diets. They found that people who ate little meat were less likely to experience depression than those who ate more meat. Following the same people for a further six years (and published in 2015) the study found that people who ate mainly plant sources had a 26 percent reduction in the risk of depression.

C Lassale *et al* (2015) conducted a meta-analysis of earlier research of Mediterranean diets. They found a one third less risk of depression with a mainly vegetarian diet compared to more omnivorous diets.

A C Tsai *et al* (2012) studied older Taiwanese people and found that people who ate the most vegetables were 62 percent less likely to develop depression than those who didn't.

A study of 50,000 people in the UK by N Ocean *et al* (2019) found that people who consumed more fruit and vegetables reported better mental wellbeing and life satisfaction than those who ate less. In fact, the more fruit and vegetables they ate, the better their reported scores.

A 2012 study by B L Beezhold *et al* looked at mood and diet. Participants in the study were randomly assigned to one of three groups. There was a group who were asked to eat meat, poultry, and fish; a group that would eat fish, but no meat; and the no meat or fish group – just fruit and vegetables. The results found that a pure plant-based diet group had a significantly boosted mood.

Other evidence looking at diet and depression comes from Asian countries. In Japan, for example, people traditionally ate mainly rice and vegetables. It was only in the 1980s and 90s that burgers and chicken take-aways became popular in the Far East and the occurrence of depression went from being comparatively rare to being much more common.

In the USA, the National Health and Nutrition Examination Survey (NHANES) between 2007 and 2014 looked at eating habits and health. They found that people who ate 21g of fibre each day were 41 percent less likely to report symptoms of depression than average. Those with a higher fibre intake each day could be up to 70 percent less likely to report depressive symptoms. The average American, the study found, ate 16g of fibre. The American Heart Association recommends 25 grams of fibre per day on a 2,000-calorie diet for adults.

The 'SMILES' trial in 2017 carried out by Felice N Jacka *et al* looked at the effect of dietary changes on existing mental illnesses over a 12-week period. One group were given social support, and a second group were put on a Mediterranean diet. At the end of the trial, the second group had a greater reduction in depressive symptoms, with one third being in remission.

So, what can we conclude from this research? If you are feeling depressed, you could start by increasing the amount of fibre that you are eating. Eat cereal for breakfast, use wholemeal bread for sandwiches, have a jacket potato for dinner. And just eat more vegetables. The other research suggests that it's best to give up meat, poultry, and fish – or cut right down and eat more fruit and vegetables. Who knows what other benefits there will be?

Resources:

Neal D Barnard, MD. Your Body in Balance: The New Science of Food, Hormones, and Health. ISBN-10: 1538747421. 2020

https://bmcmedicine.biomedcentral.com/articles/10.1186/s12916-017-0791-y

Going to Vagus

A look at the vagus nerve and vagal tone and its important to our health and wellbeing.

The word "vagus" in Latin means "wandering", and the vagus nerve is called that because it wanders round the body. It originates in the brainstem (it's the 10th cranial nerve) and goes to the lungs, heart, gut, liver, salivary glands, and kidneys. The vagus nerve affects heart rate, breathing rate, gastrointestinal peristalsis, sweating, detoxification, and much else. It also controls a few skeletal muscles, including muscles responsible for swallowing, as well as phonation (producing sound through vibration) and speech.

Let's quickly look at the nervous system. You have the central nervous system (the brain and spinal cord) and the peripheral nervous system (the other nerves). The peripheral nervous system is divided into sensory (afferent) nerves (which tell you what's going on) and motor (efferent) nerves (which tell muscles and glands etc to do something). The motor nerves are divided into somatic (voluntary) nerves (the things you can control) and autonomic (involuntary) nerves (the things you can't control). And autonomic nerves are divided into sympathetic nerves (fight-or-flight) and parasympathetic nerves (rest-and-digest). The enteric nervous system, which supplies the GI tract, is a quasi-autonomous part of autonomic nervous system.

The vagus nerve is part sensory and part motor. It has many branches, and is the longest nerve in the autonomic nervous system. It has primitive unmyelinated branches and newer myelinated branches.

The body's immune system is there to protect the body from invading bacteria etc. For example, a cut finger will result in white cells in the blood coming to surround the invading bacteria. It also results in some swelling around the cut. And this swelling – inflammation – is associated with immune responses. Inflammation can also result from stress, bereavement, poverty, debt, social isolation, maltreatment as a child, being overweight, and other one-off stressful events. Cytokines are released during inflammation. They are small proteins used in cell signalling. They can cause damage to cells around the site of the inflammation. The vagus nerve can sense cytokine levels and signal the brain. The brain will then use the vagus nerve to tell the spleen to produce fewer white cells and reduce the cytokine levels. This is called the inflammatory reflex. The cholinergic anti-inflammatory pathway is the name given to the efferent, or motor arm of the inflammatory reflex, Vagal tone effects how well inflammation is reduced. This is important because inflammation is associated with illnesses such as allergy, asthma, autoimmune diseases, coeliac disease, glomerulonephritis, hepatitis, inflammatory bowel disease, and others.

So, the vagus nerve, and therefore vagal tone, are important in controlling inflammation and inflammatory diseases.

The vagus nerve also plays a big part in, what's called, the Gut-Brain Axis (GBA). This is a communication network in both directions between the brain and the gut. Eighty percent of the information transmitted by the vagus nerve flows from the body to the brain (afferent nerve fibres). Whereas twenty percent of the vagus nerve is efferent, which means the signal are transmitted from the brain to the body.

Evidence from animal studies suggests that gut microorganisms (the biome) can activate the vagus nerve and that such activation plays a critical role in mediating effects on the brain and behaviour. At the same time, our mind (our thoughts and emotions) affects our gut health and microbiome through the vagus nerve. Stress inhibits the signals sent through the vagus nerve and causes gastrointestinal problems. Poor vagus nerve function explains why stress: suppresses stomach acid and digestive enzyme production, increases gut permeability (leaky gut), reduces the Migrating Motor Complex (a pattern of electrical activity observed in the gastrointestinal tract in a regular cycle during fasting), which is thought to be the main cause of small intestine bacterial overgrowth (SIBO), a disorder of excessive bacterial growth in the small intestine that is associated with symptoms such as nausea, bloating, vomiting, diarrhoea, malnutrition, weight loss, and malabsorption.

When immune cells first detect the presence of pathogenic or toxic agents in the gut, the vagus nerve stimulates the thymus (found in children) and the spleen to increase their activity. At the same time, a stress response and the sympathetic nerves are activated, which causes the release of the neurotransmitter noradrenalin. This in turn causes the immune system to be highly reactive to the threat. When the threat has been dealt with, the vagus nerve turns off the unneeded immune response. The vagus nerve sends out the neurotransmitter acetylcholine in the gut and other parts of the body.

Obviously, if vagal tone is low, then the immune response will not be completely turned off. Vagal tone refers to activity of the vagus nerve. The more active it is, the better the vagal tone.

How can you tell if your vagal tone is low? Apart from feeling ill, here are some examples:

- You have difficulty speaking or lose your voice
- Your voice is hoarse or wheezy
- You have trouble drinking liquids
- You lose your gag reflex
- You have abnormal heart rate and/or blood pressure
- You have decreased stomach acid and/or digestive enzymes
- You have abdominal bloating or pain
- Your bowel transit time is less than 10 hours or more than 20 hours (often resulting in diarrhoea or constipation)
- You have gastroparesis (the stomach cannot empty itself of food in a normal fashion).

How can you stimulate and improve your vagal tone? Bottom-up approaches include:

- Yoga / tai chi / qigong – 15-20 minutes is sufficient to get the benefits provided this is done regularly. Remember that breath is more important than the movement in these practices.

- Deep slow breathing – a deep diaphragmatic belly breath followed by a long slow exhale causes the vagus nerve to release acetylcholine into muscles, which has a calming effect and can improves performance where calmness is important.
- Singing/humming/chanting – it must be enthusiastic enough to cause your voice box to vibrate. It can be done in your car.
- Gargling – it must be vigorous, 2 minutes at a time.
- Cold showers – start with 5-10 seconds at the end of a shower and build up to a few minutes. 1-2 minutes is very good. In fact, any exposure to cold can be good.
- Gag reflex – stimulate the reflex by touching the back of your tongue or your soft palate with your toothbrush. The reflex must be strong.
- Sunlight – expose your skin to sunlight first thing in the morning and/or during the day)
- Walking in nature – for at least 20-30 minutes at a time.
- Sleeping on your side.

Top-down approaches include:

- Meditation
- Psychotherapy
- Positive social interaction – the emphasis being on 'positive' and feeling safe.
- Expressing gratitude
- Craniosacral therapy
- Biofeedback
- Laughter
- Prayer
- Exercise.

Clearly the vagus nerve has an important role in the body in terms of health. So, it's important that clients are as healthy as possible in order to make the other changes they want in their lives.

Why do people do what they do?

A look at all the factors affecting how a person thinks, behaves, and feels.

We see clients who want to change the way they feel – they're stuck, depressed, etc. We see clients who want to change how they behave – they want to become non-smokers or be able to give that presentation. And we see clients who want to change the way they think – they want to stop needing to perform OCD rituals or they want to get over a phobia. A person's thoughts, feelings, and behaviour are affected by more than just what happens in their brain. So, I thought I'd look at why people do what they do.

All we ever see is what people look like, what they do, and how they react to what's going on. From that, we try to infer their mood and we attempt to predict how they will behave in future. We might even guess at what they were thinking when they acted in the way they did. And, quite possibly, we'll try to work out what their beliefs might be that led them to be the way they were. In this article, I want to look at the two-way links between the brain and body affecting how a person thinks, feels, and behaves.

So, what factors do make people behave, feel, and even think the way they do?

Let's start at the beginning. People start off the way they are because of their genes – a random mixture from both parents. In addition to how the genes are expressed, there are also the effects of epigenetics. This is where the expression of a gene or genes is altered, but the basic genetic code doesn't change. As scientists discover more about genes, we find some that definitely have a specific impact on a person – eg genes for particular diseases – and there are suggested associations between genes and certain things, eg addiction or depression.

Do genes control how happy a person will be? Sonja Lyubomirsky, author of *The How of Happiness* suggests that 50 percent of happiness is genetically predetermined, while 10% is due to life circumstances, and 40 percent is the result of your own personal outlook (see Figure 1). This figure might apply to more than just happiness.

The second big thing that affects a person is their environment. This is the surroundings or conditions in which a person lives or operates.

The effects of genes and environment led to the nature-nurture debate that went on for many years. And this fits Lyubomirsky's ideas about happiness levels.

We also know about US soldiers in the Vietnam war who took addictive opium while in Vietnam, and how the majority of them just stopped when they returned to their families in the USA. Their environment changed and so did their behaviour.

We also know that a person's childhood can affect how they think and behave as an adult, and also affect their health as an adult and their life expectancy. It's also during childhood that many people's core values and beliefs are created (or indoctrinated – depending on your beliefs).

We help clients who experience butterflies or worse with nerves or excitement at the thought of giving a presentation and many other activities. These can be affected by

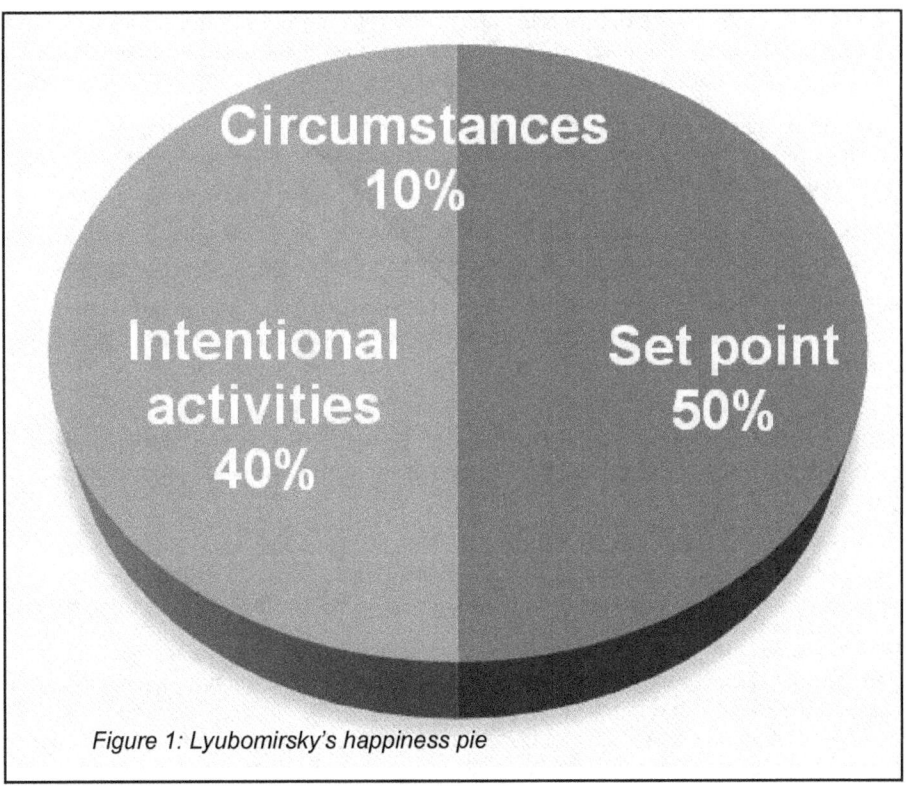

Figure 1: Lyubomirsky's happiness pie

their expectations of what will happen – or their lack of information about what will happen.

And we see people looking for help with IBS (Irritable Bowel Syndrome) and other gut-related issues. So, let's move our focus to our GI tract and the millions of bacteria etc that live in it – our gut microbiome. And let's look at the two-way link between the brain and gut – the Gut-Brain Axis (GBA). Our gut has to digest food (ie make and secrete enzymes). It has to absorb food. It has to squeeze food through itself (peristalsis). And it has to defeat invading bacteria. It has to ensure the cells in the lining keep the bad stuff out and let the good stuff through. It has to not have too thin a layer or too many gaps between cells. And it has to produce a layer of mucous over the top of this lining. And your microbiome helps – it does use some of your food, but it does create useful products. If you don't have a microbiome, eg through taking too many antibiotics, you'll not be very healthy.

What you eat can affects your biome. Studies have shown that gut microbes can affect behaviour and even emotions (like depression). A Belgian study found two kinds of microbe (Coprococcus and Dialister) were missing from the microbiomes of their depressed subjects, but not from those with a high quality of life.

Compounds made by the gut microbiome, eg Short-Chain Fatty Acids (SCFAs), such as butyric acid, propionic acid, and acetic acid, are able to stimulate sympathetic nerves, mucosal serotonin release, and to influence memory and the learning process.

Vagal nerve activity and faecal microbiota transplants can be used to influence mental health. The gut microbiome plays a facilitating role between the stress response, inflammation and depression, and anxiety.

Intestinal permeability (ie leaky gut) occurs with chronic low-grade inflammation, which happens more often in disorders such as anxiety or depression.

It is estimated that 90 percent of the body's serotonin is made in the digestive tract. In fact, altered levels of gut serotonin have been linked to diseases such as irritable bowel syndrome, cardiovascular disease, and osteoporosis.

Stress can affect the composition of, and the total amount of, biome in a person's gut. The biome can be directly affected by neurons, immune cells, and enterochromaffin cells (neuroendocrine cells found in the gastric glands, that aid in the production of gastric acid through the release of histamine).

The brain also modulates gut functions such as: motility; the secretion of acid, bicarbonates, and mucus; intestinal fluid handling; and mucosal immune response. These maintain the mucus layer and biofilm where individual groups of bacteria grow.

Plus, the brain may affect the biome composition and function by changing intestinal permeability, allowing bacterial antigens to penetrate the epithelium and stimulate an immune response in the mucosa (mucous membrane). So, through the autonomic nervous system, the brain modulates immune function, which can increase epithelial permeability to bacteria, which facilitates their access to immune cells.

Changes in the composition of the gut flora due to diet, drugs, or disease correlate with changes in levels of circulating cytokines, some of which can affect brain function. Cytokines are small proteins that affect the behaviour of cells around them. They are especially important in the immune system.

Dysbiosis describes an imbalance in the gut microbiome, which can lead to an overgrowth of opportunistic gastrointestinal microorganisms, a reduction in the production of short-chain fatty acids, and reduced resistance to intestinal pathogens. The depletion of beneficial bacteria that occurs during dysbiosis is also understood to impair the immune system and trigger inflammatory disorders such as inflammatory bowel disease.

Lastly on this, the gut has its own nervous system – called the enteric nervous system.

The vagus nerve wanders round the body and contains nerves sending messages from the brain, and nerves sending messages to the brain. It affects how the gut behaves (trying to keep everything well – a homeostatic role), but it's also part of the Gut-Brain Axis, sending messages from the gut to the brain, which then impact on mood.

Most of the body's immune system works on the gut! When our immune system identifies an invader, it releases cytokines, and the body is protected by inflammation as white cells attack the invading organisms. However, some people have low levels of inflammation all the time.

It's worth noting that the brain and the immune system have two-way links.

Stress can cause inflammation. Stressful events include bereavement, poverty, debt, social isolation, maltreatment as a child. Being overweight causes inflammation. Even public speaking can increase inflammation!

Most inflammation occurs on the inside of the body, which people can't see. One of the ways that the body naturally keeps the levels of inflammation down is through the vagus nerve. It controls the inflammatory reflex.

The inflammatory reflex is a neural circuit that regulates the immune response to injury and invasion. If cytokine levels in the body rise, the vagus nerve will detect the change and send a message to the brain. A signal is then sent down the vagus nerve to the spleen, acting on macrophages (white blood cells) to reduce the cytokine level. Cytokines can cause collateral damage to the body's cells near them. Interestingly, vagal nerve stimulation can reduce inflammation. Increased vagal signalling inhibits inflammation and prevents organ damage.

The health and fitness of the vagus nerve is called vagal tone. A high vagal tone equates to a better capacity to keep inflammation down.

Other factors associated with increased inflammation include: obesity; sedentary lifestyle; disordered sleep; emotional and physical trauma; medical illnesses such as cardiovascular disease, diabetes, cancer, and autoimmune; infections (including exposure to unsanitary living conditions and poor hygiene); medical treatment such as surgery, chemotherapy, or radiation; and antidepressant treatment resistance.

The body's natural immune response can trigger oxidative stress temporarily. This type of oxidative stress causes mild inflammation that goes away after the immune system fights off an infection or repairs an injury. Oxidative stress refers to an imbalance of free radicals and antioxidants in the body, which can lead to cell and tissue damage. Free radicals are molecules with one or more unpaired electron that are very reactive. Antioxidants are substances that neutralize or remove free radicals by donating an electron. The neutralizing effect of antioxidants helps protect the body from oxidative stress. Examples of antioxidants include vitamins A, C, and E.

There's lots of research evidence linking inflammation with depression.

The blood-brain barrier protects the brain. We used to think it was impenetrable.

The neuroimmune system is composed primarily of glial cells and mast cells (a type of white blood cell). During a neuroimmune response, cytokines send inflammatory signals across the blood-brain barrier, which activates microglial cells in the brain, which then release more cytokines. As a consequence, this can kill neurons or shrink them, reduce the number of synaptic connections, and the synaptic supply of neurotransmitters can be disrupted. And it can block the regenerative process that would create new cells. Tryptophan is a serotonin precursor. Microglial cells can instruct nerve cells to make other end products such as kynurenine. This makes less serotonin available in the brain and kynurenine (and other alternative end products) are toxic. So, inflammation in the body can have a negative impact on the brain – perhaps, leading to depression.

What else can affect a person? Well, there's sleep, exercise, and nutrition.

But what happens if you don't get enough sleep? Insomnia:

- Is linked with depression and anxiety
- Makes you forgetful
- Impairs your judgement
- Cognitively impairs your thinking
- Causes accidents, eg falling asleep at the wheel
- Is linked to health issues (heart attack, stroke, diabetes)
- Kills sex drive
- Ages your skin
- Can cause weight gain
- Increases the risk of death.

If you do get enough sleep, it:

- Reduces stress
- Reduces the risk of depression
- Makes you more alert
- Improves your memory
- Cleans up your brain
- Makes you cleverer
- Helps your body repair itself
- Reduces inflammation
- Keeps your heart healthy
- May prevent cancer
- May help you lose weight.

Research shows there are links between insomnia and anxiety.

The benefits of exercise (movement) are that it:

- Creates new brain cells (neurogenesis) using BDNF (brain-derived neurotrophic factor).
- Improves your mental health and mood
- Helps keep you thinking and learning

- Improves your sleep
- Strengthens bones and muscles
- Reduces your risk of heart disease
- Help you manage your blood sugar and insulin levels
- Increases your chances of living longer
- Improves your sexual health
- Reduces the risk of some cancers.

Nutrition helps with your biome and helps your GI (gastrointestinal) tract.

Good things to eat include: all the food groups (carbohydrates, fats, proteins, vitamins, minerals, and water) in moderation, prebiotics and probiotics, anti-inflammatory foods (eg fruits, greens, nuts, and grains), and foods containing antioxidants (eg dark chocolate, whole cereal, fruit, and vegetables).

Food tips for a good mood include: eat breakfast, stay hydrated, eat regularly, eat the right kind of fats (not trans fats), and avoid processed foods.

All of these impact on your body and your brain, and therefore how you feel, think, and behave.

As you go through life, you develop your model of the world. These experiences will usually be similar but different to your friends and family and this will lead to your beliefs and your values. These will affect how you behave and how you feel about things. Of course, your memories of an experience will change every time you recall that memory.

What else can affect how we feel? If we are suffering pain, that can have a big negative impact. Our hormone levels can affect us (for example menopausal women, or if our body is flooded with adrenalin). If we tend to brood (overusing that right pre-frontal cortex), that can impact on how we feel.

If we control our breathing (for example 7-11 breathing, square breathing), that can have a positive impact on how we feel. Diaphragmatic breathing with slow exhalations stimulates the vagus nerve and slows our heart rate and reduces our blood pressure.

Laughter can have a huge positive effect on how we feel.

Meditation, eg the Loving-kindness meditation can help us to feel better and then behave in a more positive way. Here's how...

Say three times: "May I be filled with loving kindness, and be well, peaceful and at ease, happy, and free of suffering".

Then say three times for a loved one: "May [loved one] be filled with loving kindness, and be well, peaceful and at ease, happy, and free of suffering".

Lastly, say three times for each: a neutral person, a difficult person, and all sentient beings. May [neutral person] be filled with loving kindness, and be well, peaceful and at ease, happy, and free of suffering."

This constitutes one cycle. You can do as many or as few cycles as you wish.

Doing things that make us happy is good, but, often, what we think will make us happy doesn't. Hedonistic happiness includes things like eating chocolate and going to the pub every Friday. Hedonic adaptation results in us not enjoying them as much as we expected. Eudaimonic happiness (or wellbeing) refers to the subjective experiences associated with living a life of virtue in pursuit of human excellence. What that really means is experiences are usually more enjoyable than things.

Events can have a big impact on where we are on the happy-sad scale. Evidence shows that if you have a big win (or a big loss) in life, typically, after a while you are back to your original level of happiness. In fact, how we interpret events has a bigger impact than the event itself. So, it's estimated that 90 percent of the impact of an event is how we interpret it, and only 10 percent is the event itself.

Pressure can impact us negatively. This is where someone (our boss) demands work is done by a particular time. This causes negative stress in our life. Too much pressure, whether real or imagined, can lead some people to thoughts of suicide.

Other people's actions and reactions to what we say and do can not only impact on how we feel and how we behave, it can also impact on our values and beliefs. So, spend time with positive people.

NLP has a model of what goes on in a person's head and this affects how successful an attempt to make a change will be. Robert Dilts came up with a model of what, he called, logical levels (see Figure 2).

At the bottom is environment. This is like the opium addicts in the Vietnam war. Changing their environment (ie going home) changed their behaviour. Above environment is behaviour, above that is capabilities – you can only do what you're

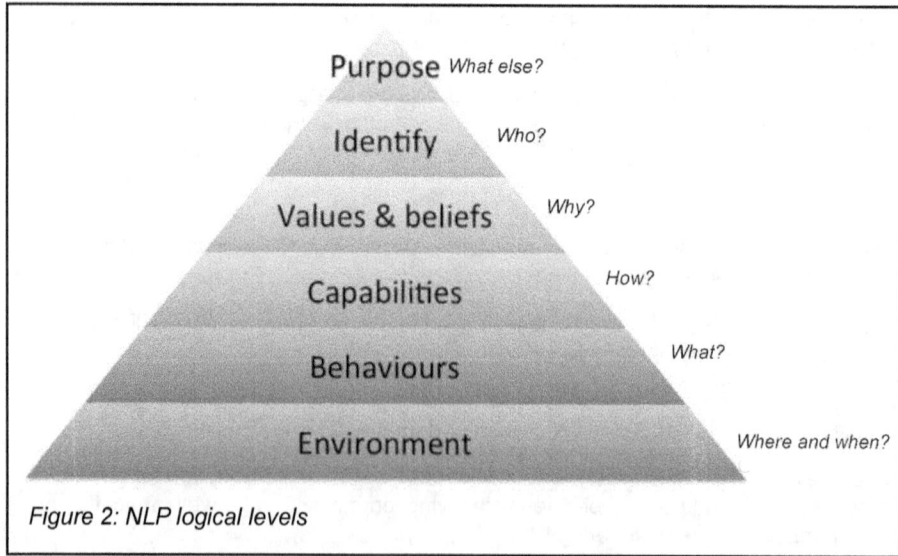

Figure 2: NLP logical levels

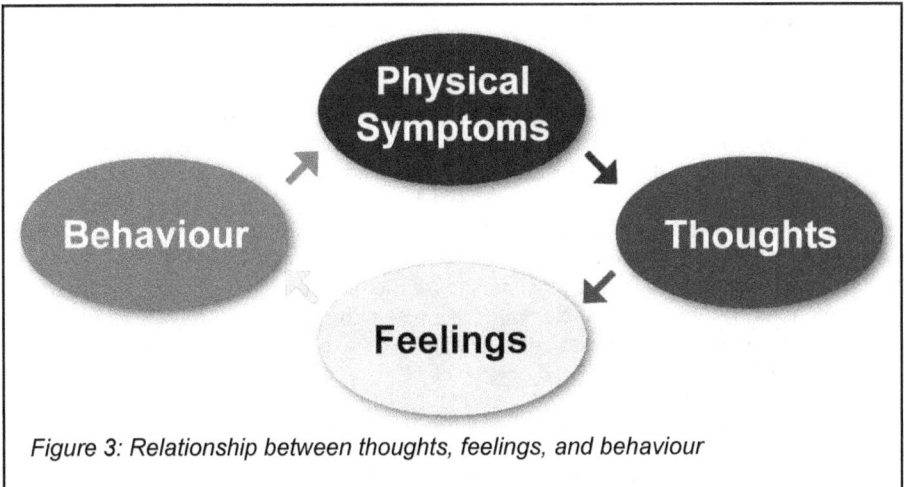

Figure 3: Relationship between thoughts, feelings, and behaviour

capable of doing. Above that are beliefs and values, above that is identity, which, I guess, used to be more important where someone was identified by their job, for example doctor, colonel, etc. Above that is put mission or purpose, or, sometimes, spirituality.

Other things people talk about in terms of getting themselves to do something are motivation, resilience (if other things or people are trying to stop them), and willpower (or grit when the going gets tough).

And if we want to be positive about things, then we need to have positive thoughts, positive actions, positive interactions, and a purpose to our life. If we don't have enough of any of those, this can lead to negative emotions and thoughts.

I also want to mention mindset. People with a fixed mindset think that their positive characteristics are fixed and there's nothing they can do about them, so they will behave in ways to hide their failures and make the most of their successes. People with a growth mindset know that they will be able to work hard and improve throughout their life. So, they will take risks with tasks because they treat failure as a learning exercise.

If you are feeling anxious, then this can lead to behaviours like OCD, fears and phobias, and panic attacks. It can make pain hurt more.

And thoughts, feelings, behaviour, and physical reactions (eg blushing) are linked, so a change in one can cause a change in the others (see Figure 3).

Just a thought about emotions. Lisa Feldman Barrett suggests that emotions are constructed in brain circuits – rather than just appearing in the amygdala.

Anyway, that's a very quick run through the different factors affecting why people do what they do, and that's without even looking at the things we already know about the brain.

Why aren't I happier?

A look at what clients think will make them happy and what actually does.

I can remember, a few years ago, talking to someone who assured me that once they got their hands on the new iPhone that was coming out next week, they would be the happiest person in the world. I saw them just over a week later and they proudly showed off their new phone (I did say it was a few years ago), and told me the story of how far they had to travel to get to the shop and how long the queue was etc. But they were definitely happy. Proof, if proof were needed, that possessions make you happy! Except that I saw them again, a few times, over the next few months, and, by the end of that period, they were no happier with their new phone than they had been before they'd got it. Oh dear, proof that possessions don't make you happier – or if they do, it's only for a short while.

Let's try another example: let's think about life in the 1940s. I bet people are happier now with central heating, mobile phones, and 60-inch TV sets? Would it surprise you to find that we're not! In her 2007 book, *The How of Happiness: A New Approach to Getting the Life You Want,* Sonja Lyubomirsky looked at how life satisfaction in the 1940s compared with more recent times. And, guess what? She found that we're not happier now. Similarly, David G Myers in his 2000 book, *The American Paradox: Spiritual Hunger in an Age of Plenty,* found that "our becoming much better off over the last four decades has not been accompanied by one iota of increased subjective wellbeing".

As hypnotherapists, we see a lot of people for weight loss. They expect to be happier when they've dropped a dress size (or whatever). Yet, a 2014 study by Jackson *et al* found that weight loss didn't necessarily make people feel any happier. Or how about going for cosmetic surgery – wouldn't that solve all their problems and make them completely happy? A 2012 study by von Soest *et al* looking at women in Norway following cosmetic surgery and its effects on psychological factors and mental health found that plastic surgery just didn't seem to alleviate mental health problems.

What about finding true love – isn't that the route to perfect happiness. The answer is yes and no. A 2003 study by Lucas *et al* found that people get happier as they get nearer to their wedding, but once they've got used to being married (about 2 years' later) their happiness returns to its previous baseline value.

But I bet having plenty of money makes you happier? That's what Diener & Oishi (2000) set out to investigate. They looked at life satisfaction and income in different countries. They didn't find a strong relationship between the two, suggesting that income doesn't, necessarily, make you happier.

In 2010, Kahneman & Deaton looked at the effects of a high income. They found that money improves a person's evaluation of their life but not their emotional wellbeing. Once you reach a certain income, more money won't make you happier.

The trouble with money is that it is comparative. So, if I told you that I was giving you a million pound, you'd be very happy. That is until you heard I was giving everyone else two million! Sonja Lyubormirsky also showed in her book that a person's salary goals rise as their salary rises. Tantalizingly, you never reach the salary you think you deserve.

You probably think that those findings may be true for other people, but not for you. Well, again, research has found that people aren't very good at predicting by how much things will make them happier or by how much bad things will make them unhappy. Let's suppose that you applied for a job that you wanted, but you didn't get it. I imagine that you'd feel disappointed and upset. A study by Gilbert *et al* (1998) found that people just didn't feel as bad as they expected when they were turned down for a job.

So, we can conclude from this that what you expect will make you happy doesn't – or won't for very long. You never get paid enough. And people aren't very good at predicting how happy or sad they're going to feel after an event occurs. It's worth pointing this out to unhappy clients.

Some people think that how happy they are (their happiness level) is predetermined by their genes. However, Sonja Lyubomirsky (again) in *The How of Happiness* suggests that 50 percent of happiness is genetically predetermined, while 10% is due to life circumstances, and 40 percent is the result of a person's personal outlook. So, we can conclude, people do have control over how happy they are.

Why aren't we very good at predicting our happiness levels? It seems that most people predict their happiness in terms of absolutes – if I get x amount of salary, if I get x car or phone, if I get a house with x number of bedrooms, etc, I will be 10 on a scale of 1 to 10. But, it seems, in real life, people compare themselves to other people. Their happiness is relative to other people's happiness. My four-bedroom house is not that good when my work colleague has a five-bedroom house! They have a relative reference point rather than (as they think) an absolute reference point.

Let's look at a real-life example. If I win a medal at the Olympics, I would probably predict that I will be happier than those athletes who don't win any medals. That sounds sensible, doesn't it? I would predict that the gold medal winner is the happiest, the silver medallist slightly less happy, and the bronze medallist still very happy but less so than the silver medallist. A 1995 study by Medvec *et al* looked at pictures of winners on the podium. As expected, gold medal winners are very happy. Perhaps surprisingly, silver medal winners seem almost unhappy. And that's because, it's suggested, their reference point is the gold medallist – who was better than them. The bronze medallist looks happier than the silver medallist. And that is because their reference point is the rest of the competitors. They are happy because they were better than everyone else.

Clark and Oswald in 1996 surveyed 5,000 British workers looking at a person's job satisfaction. They found something similar. It turned out that where a person's co-workers earned more money than they did, that person became less satisfied with their job. It didn't matter, absolutely, how much they earned, it was the relative reference point (their co-workers) that mattered.

In 1997 Solnick and Hemenway looked at relative reference points with students. The students were given a choice. Option 1 was a job where they earned $50,000, and everyone else on the same grade earned $25,000. Option 2 was a job where they earned $100,000, but everyone else on the same grade earned $250,000. Over 50 percent chose option 1 – the lower salary – because of the comparison.

Similarly, a study by Clark in 2003 found that unemployed people were happier if the unemployment rate was high in their area rather than if they lived in an area of low unemployment. Another example of people using a relative reference point.

The worrying thing about using a relative reference point is that we compare ourselves against that reference. So, you might predict, that if we watch TV programmes full of good-looking healthy people, we will feel less good about ourselves. And if we compare ourselves with our neighbours, we end up 'keeping up with the Joneses'.

O'Guinn and Shrum in 1997 found that TV programmes featuring products and activities associated with an affluent lifestyle act as a harmful social comparison skewing a person's perception of other people's wealth and their own wealth. Similarly, Juliet B Schor's 1999 book, *The Overspent American: Why We Want What We Don't Need*, suggests that watching TV acts as a harmful social comparison and increases the amount a person spends each week. Or, suppose your neighbours win the lottery and buy a new car, what are you likely to do? The answer is also to buy a new car! Kuhn *et al* (2011) looked at the Dutch Postcode Lottery and found that people living next door to lottery winners are more likely to buy a new car. It's another example of social comparison influencing our spending. And there are plenty of other similar results from research.

Social comparison on social media can also have negative effect. In 2014, Vogel *et al* looked at social comparison on social media and a person's self-esteem. They found that comparing ourselves to others on social media lowers our self-esteem. They also manipulated the Facebook feed to feature people who were worse off than the subjects. Even that didn't lead to much higher self-esteem ratings.

So, it's worth pointing out to clients that their unhappiness may be due to inappropriate comparisons with other people.

I mentioned earlier that people get used to good things and so don't stay happier than they were previously. This is called hedonic adaptation. This applies to lottery wins, getting married, pay rises, and most other things. Daniel Gilbert in his 2007 book, *Stumbling on Happiness*, says that "wonderful things are especially wonderful the first time they happen, but their wonderfulness wanes with repetition".

As mentioned previously, people are bad at predicting how happy or sad events are going to make them. They tend to overestimate how happy they will be following a good event and overestimate how bad they will feel following a bad event. Research indicates that people are more resilient when facing bad things and less excited about good ones. Or their estimates are just completely off!

Gilbert *et al* (1998) found that people are generally unaware of their psychological immune system, which is why they tend to overestimate their emotional reactions to negative events. Similarly, Eastwick *et al* (2008) found that people mis-predict how they will feel if they break up with a significant other. They think they will feel much worse than we actually do. Ayton *et al* (2007) suggested that a person's affective forecasts (predicting their emotional response given a certain outcome) are too extreme. Worryingly, the research also found that people don't seem able to predict their response with any greater accuracy even when they have previous experience of that emotional event!

One reason given for why people overestimate the effect of an event on their happiness is that they focus on that single event and don't include all the other things going on in their life. Again, Daniel Gilbert's 2007 book, *Stumbling on Happiness*, offers some suggestions as to why people mis-predict. He suggests two cognitive biases, focalism, and immune neglect. Focalism is the tendency of people to think about just the one event and forget about the other things that happen in their life. Immune neglect is a person's unawareness of their tendency to adapt to and cope with negative events. Dunn *et al* (2002) wrote that our predictions are worse for negative events. They suggested that when a person thinks about the future, they tend to focus on the wrong features and overestimate their importance.

So, what can people do to make themselves happier? Clearly, people don't seem to know what will make them happy and they can't predict how happy (or sad) anything going to make them. The good news is that there are a number of things people can do to make themselves happier. Firstly, they can spend their money on experiences rather than things. Boven & Gilovich (2003) found that it's better to *do* than to *have*, ie experiences make people happier. And Kumar *et al* (2014) found that experiences have a longer-lasting effect on happiness. However, Pchelin & Howell (2014) warned us that when people are looking at future purchases they are more likely to value material purchases over experiential purchase, even though when they look at past purchases they're more likely to value experiences over material goods. Howell & Hill (2009) advised that experiential purchases make a person feel more alive and they are less susceptible to social comparisons.

Another technique to make people happy is savouring. Savouring is the use of thoughts and actions to increase the intensity, duration, and appreciation of positive experiences and emotions. For example, Jose *et al* (2012) found that savouring positive experiences makes a person happier. And Lyubomirsky *et al* (2006) found that thinking about life's positive moments makes a person happier. Surprisingly, they also found that writing about life's negative moments made a person happier too. Koo *et al* (2008) found that thinking about how something good in your life might not have happened if things had been different actually makes you happier. Kurtz (2008) looked at whether focusing on the imminent ending of a positive life experience can lead to increased enjoyment. The research found that thinking about an experience coming to an end can enhance a person's present enjoyment of that experience.

Gratitude is another way of increasing a person's happiness. Emmons *et al* (2003) found that if a person counts their blessings (the good things in their life), they become happier. A 2015 study by Barton *et al* looked at financial distress and marital quality. It found that being grateful can help people through difficult times. Grant & Gino in 2010 found that receiving gratitude makes people feel valued and motivates them to be more generous.

Strangely, people enjoy things more when they are interrupted. It seems to disrupt the normal process of hedonic adaptation. For example, Nelson & Meyvis (2008) found that, despite not wanting them, people actually enjoy positive experiences more when there are breaks. And Nelson *et al* (2009) found that adverts actually made the experience of watching TV more positive. Don't tell Netflix!

Martin Seligman, who came up with the idea of positive psychology, also came up with idea of signature strengths in his book, *Authentic Happiness*. Signature strengths are desires or disposition to act or feelings that seem to lead to recognizable excellence or instances of flourishing. They are the character strengths that are most essential to who we are. And Seligman identified 24 of them. They tend to be morally valued in most moral systems of the world. Seligman *et al* (2005) found that when people used their top signature strengths in a new and different way every day for one week, it had an enduring impact on their happiness. Similarly, Lavy & Littman-Ovadia (2017) found that people who use signature strengths at work are more productive and more satisfied with their job. While Harzer & Ruch (2012) found that people enjoy work more and think of work as a calling when they use around four of their signature strengths at work. So, that's another way that people can be happier.

What other things can you do to make you happy? Being kind makes you happy. That's what Otake *et al* found in 2006. Happy people become happier through kindness – the kinder you are, the happier you become. Similarly, Lyubomirsky in 2005 found that people who intentionally carried out random acts of kindness made themself happier. Elizabeth Dunn wrote in her book, *Happy Money: The Science of Happier Spending*, that money can buy happiness if you spend it on the right things, such as spending money on others rather than yourself. This matched Dunn *et al*'s 2008 findings that spending money on other people promotes happiness and makes you feel good.

We always talk about the 4Ps to clients, and there's a lot of evidence showing that the third P – positive interaction is good for people. In 2000, David Myers found that having strong social ties makes people healthier. In 2002 Diener & Seligman wrote that being social/having strong social ties makes people happier. Epley & Schroeder (2014) found something that at first glance seems completely counterintuitive. Talking to people on the bus or the train is better than sitting in solitude! In their experiment, two people came into a sort of waiting room. One subject was instructed to talk, not talk, or just do what they would normally do. In every case where people talked, they felt better than where they didn't. So, it's worth striking up a conversation with strangers – both of you get a happiness boost after the chat. The findings of Boothby *et al* (2014) are not surprising from our past experience that if we share an experience with another person, then the experience is amplified and enjoyed more by the people taking part.

What makes you happier – having more time or more money? Whillans *et al* (2016) showed people two examples. There was Tina, who values her time more than money. She's willing to sacrifice money to have more free time. She'd rather work fewer hours. Then there's Maggie, who values her money more than her time. She's willing to sacrifice lots of time to get more money. She'd work more hours to earn more money rather than take time off. The subjects were asked which example they were more like, and then how happy they were. Researchers found that people who prioritize time over money, ie were more like Tina, were typically happier than people who prioritize money over time. Hershfield *et al* (2016) also found that people who choose time over money are happier, but they also found around 70 percent of people choose money over time, and only 30 percent choose time over money. Cassie Moligner (2010) found that thinking about time makes a person happier than thinking about money, however thinking about time boosts their motivation to socialize – which is associated with greater happiness.

Armed with this information, it should be possible to help your clients to think about what they want and help them to do things that will make them happier.

References:

Ayton *et al* (2007). Affective forecasting: Why can't people predict their emotions? Thinking & Reasoning, 13, 62-80.

Barton *et al* (2015). Linking financial distress to marital quality: The intermediary roles of demand/withdraw and spousal gratitude expressions. Personal Relationships, 22, 536–549.

Boothby *et al* (2014). Shared experiences are amplified. Psychological Science, 25(12), 2209-2216.

Boven & Gilovich (2003). To Do or to Have? That Is the Question. Journal of Personality and Social Psychology, 85(6), 1193–1202.

Clark (2003). Unemployment as a social norm: Psychological evidence from panel data. Journal of Labor Economics, 21(2), 323-351.

Clark and Oswald (1996). Satisfaction and comparison income. Journal of Public Economics, 61(3) 359-381.

Diener & Oishi (2000). Money and happiness: Income and subjective well-being across nations. Culture and Subjective Well-Being, Cambridge, MA: MIT Press.

Diener & Seligman (2002). Very happy people. Psychological science, 13(1), 81-84.

Dunn (2014). Happy Money: The Science of Happier Spending. New York, NY: Simon & Schuster.

Dunn *et al* (2002). Location, location, location: The misprediction of satisfaction in housing lotteries. Personality and Social Psychology Bulletin, 29(11),1421-1432.

Dunn *et al* (2008). Spending money on others promotes happiness. Science,319 (5870), 1687-1688.

Eastwick *et al* (2008). Mispredicting distress following romantic breakup: Revealing the time course of the affective forecasting error. Journal of Experimental Social Psychology, 44, 800-807.

Emmons *et al* (2003). Counting blessings versus burdens: An experimental investigation of gratitude and subjective well-being in daily life. Journal of personality and social psychology, 84(2), 377.

Epley & Schroeder (2014). Mistakenly seeking solitude. Journal of Experimental Psychology: General, 143(5), 1980.

Gilbert (2007). Stumbling on Happiness. Gilbert (2007) New York; NY: Vintage Books.

Gilbert *et al* (1998). Immune neglect: A source of durability bias in affective forecasting. Journal of Personality and Social Psychology, 75, 617-638.

Grant & Gino (2010). A little thanks goes a long way: Explaining why gratitude expressions motivate prosocial behavior. Journal of personality and social psychology, 98(6), 946.

Harzer & Ruch (2012). When the job is a calling: The role of applying one's signature strengths at work. The Journal of Positive Psychology, 7,362-371.

Harzer & Ruch (2012). When the job is a calling: The role of applying one's signature strengths at work. The Journal of Positive Psychology, 7,362-371.

Hershfield et al (2016). People who choose time over money are happier. Social Psychological and Personality Science,7(7), 697-706.

Howell & Hill (2009). The mediators of experiential purchases: Determining the impact of psychological needs satisfaction and social comparison. The Journal of Positive Psychology, 4(6), 511-522.

Jackson et al (2014). Psychological changes following weight loss in overweight and obese adults: A prospective cohort study. PLOS, 9(8): e104552.

Jose et al (2012). Does savoring increase happiness? A daily diary study. The Journal of Positive Psychology, 7(3), 176-187.

Kahneman & Deaton (2010). High income improves evaluation of life but not emotional well-being. PNAS, 107(38), 16489-16493.

Koo et al (2008). It's a wonderful life: Mentally subtracting positive events improves people's affective states, contrary to their affective forecasts. Journal of Personality and Social Psychology, 95(5), 1217–1224.

Kuhn et al (2011). The effects of lottery prizes on winners and their neighbors: Evidence from the Dutch Postcode Lottery. American Economic Review, 101(5), 2226-2247.

Kumar et al (2014). Waiting for Merlot: Anticipatory Consumption of Experiential and Material Purchases. Psychological Science, 25(10),1924-1931.

Kurtz (2008). Looking to the future to appreciate the present: The benefits of perceived temporal scarcity. Psychological Science, 19(10), 1238-1241.

Lavy & Littman-Ovadia (2017). My better self: Using strengths at work and work productivity, organizational citizenship behavior, and satisfaction. Journal of Career Development, 44(2) 95-109

Lavy & Littman-Ovadia (2017). My better self: Using strengths at work and work productivity, organizational citizenship behavior, and satisfaction. Journal of Career Development, 44(2) 95-109

Lucas et al (2003). Reexamining Adaptation and the Set Point Model of Happiness: Reactions to Changes in Marital Status. Journal of Personality and Social Psychology, 84(3), 527-539.

Lyubomirsky (2005). Pursuing happiness: The architecture of sustainable change. Review of general psychology, 9(2), 111.

Lyubomirsky (2007). The How of Happiness: A New Approach to Getting the Life You Want. New York, NY: Penguin Books.

Lyubomirsky *et al* (2006). The costs and benefits of writing, talking, and thinking about life's triumphs and defeats. Journal of personality and social psychology, 90(4), 692.

Medvec *et al* (1995). When less is more: Counterfactual thinking and satisfaction among Olympic medalists. Journal of Personality and Social Psychology, 69(4), 603–610.

Moligner (2010). The pursuit of happiness: Time, money, and social connection. Psychological Science, Psychological Science 21(9) 1348–1354)]

Myers (2000). The American Paradox: Spiritual Hunger in an Age of Plenty. New Haven, CT: Yale University Press.

Myers (2000). The funds, friends, and faith of happy people. American psychologist, 55(1), 56.

Nelson & Meyvis (2008). Interrupted consumption: Adaptation and the disruption of hedonic experience. Journal of Marketing Research, 45(6), 654-664.

Nelson *et al* (2009). Enhancing the Television-Viewing Experience through Commercial Interruptions. Journal of Consumer Research, 36(2), 160-172.

O'Guinn and Shrum (1997). The role of television in the construction of consumer reality. Journal of Consumer Research, 23(4), 278-294.

Otake *et al* (2006). Happy people become happier through kindness: A counting kindnesses intervention. Journal of happiness studies, 7(3), 361-375.

Pchelin & Howell (2014). The hidden cost of value-seeking: People do not accurately forecast the economic benefits of experiential purchases. The Journal of Positive Psychology,9(4), 322-334.

Schor (1999). The Overspent American: Why We Want What We Don't Need. New York: NY: Harper Perennial.

Seligman (2004). Authentic Happiness: Using the New Positive Psychology to Realize Your Potential for Lasting Fulfillment. New York, NY: Simon and Schuster.

Seligman *et al* (2005). Positive Psychology Progress: Empirical Validation of Interventions. American Psychologist, 60(5):410-421

Solnick and Hemenway (1997). Is more always better?: A survey on positional concerns. Journal of Economic Behavior and Organization, 37, 373-383.

Vogel *et al* (2014). Social comparison, social media, and self-esteem. Psychology of Popular Media Culture, 3(4), 206-222.

von Soest *et al* (2012). Predictors of cosmetic surgery and its effects on psychological factors and mental health: a population-based follow-up study among Norwegian females. Psychological Medicine, 42(3), 617-626.

Whillans *et al* (2016). Valuing time over money is associated with greater happiness. Social Psychological and Personality Science, 7(3), 213-222

There's a gene for that?

A look at genetics and the work we do.

Most people seem to think that the way a person is – happy, tall, confident – is down to a mixture of genetics and environment – the old nature/nurture discussion. And, obviously, certain characteristics must be controlled by genes, because you can see that they run in families – like the Habsburg nose or porphyria in Queen Victoria's descendants.

A recent book by Robert Plomin entitled, *Blueprint: How DNA Makes Us Who We Are*, looks at recent conclusions that have been drawn from genetic research including twin studies. Plomin explains that 95 percent of a person's eye colour is controlled by genes. 80 percent of their height and 70 percent of their weight are also controlled by genes. Schizophrenia is 50 percent controlled by genes, the same as general intelligence.

The problem that Plomin highlights is with psychological conditions. He argues that these conditions shouldn't be treated as either something a person has or they don't. He suggests that this model leads to psychiatrists then looking for a cause for the condition. Plomin argues that, unlike eye colour, there are no single genes for psychological conditions. Instead, there is a bell-shaped curve, and people with psychological conditions are one end of the curve. Figure 1 shows a typical bell-shaped curve.

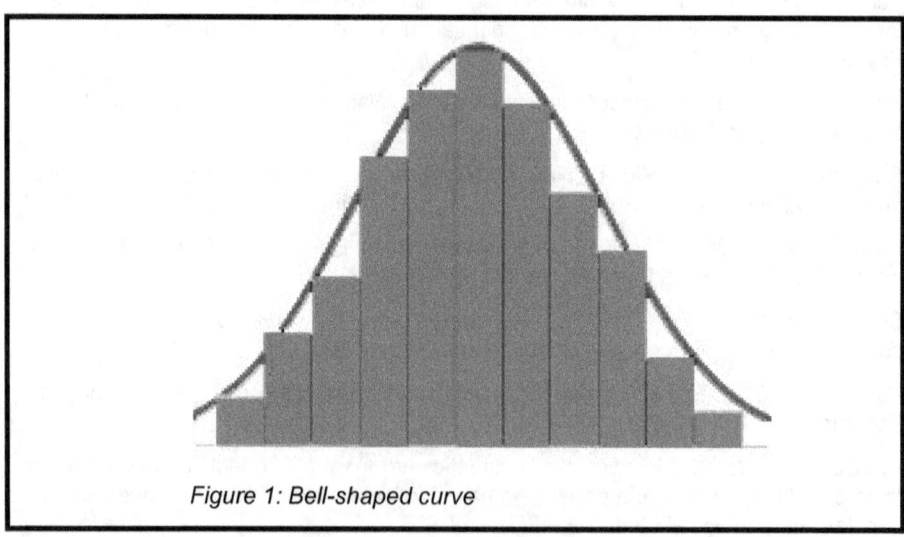

Figure 1: Bell-shaped curve

Plomin asserts that there are a large number of DNA differences that are related to a psychiatric disorder. The more of those difference that a person has, the more likely they are to experience a psychological problem. The issue, he concludes, is quantitative, not qualitative. So, if there isn't one specific gene responsible for each psychological disorder, what is there? It seems there are generalist genes, whose effects are spread across a number of conditions. Plomin cites examples where a parent has a diagnosis of depression, but the child has a diagnosis of antisocial behaviour. Surprisingly, not the same condition, which you would expect if there were to be a single gene for that condition. Plomin informs us that developmental studies have found that one condition often changes into another. And twin studies have found that generalized anxiety disorder and major depressive disorder are the same thing, genetically.

In contrast to the dozens of disorders in psychiatrists' diagnostic manuals, Plomin's research has found three broad genetic clusters. They are:

- Internalizing problems, like anxiety and depression
- Externalizing problems, like antisocial behaviour and alcohol dependency
- Psychotic experiences, like hallucinations, and includes schizophrenia, bipolar disorder, and major depression.

For the people who like to understand the technical terms used in genetics, pleiotropy is the name given to one gene that influences two or more seemingly unrelated phenotypic traits. A phenotype is an individual's observable traits, eg height, eye colour, and blood type. Polygenicity is where one particular trait is influenced by more than one gene. Traits that display a continuous distribution, such as height or skin colour, are polygenic.

A polygenic score is a correlation coefficient. A genome-wide association study (GWAS) identifies single nucleotide polymorphisms (SNPs) in the DNA that correlate with the trait of interest. The SNPs are markers only. Although they might, in some cases, suggest genomic neighbourhoods in which to search for genes that directly affect the trait, the polygenic score itself is in no sense causal. Plomin understands this and says so repeatedly in the book — yet contradicts himself several times by arguing that the scores are, in fact, causal.

A genome-wide association (GWA) study is an observational study of a genome-wide set of genetic variants in different individuals to see if any variant is associated with a trait. So, the GWA study identifies single nucleotide polymorphisms (SNPs) in the DNA that correlate with the trait of interest. The SNPs are markers only. This has revealed tens of thousands of genes that can have an effect, and the average effect size is 0.01 percent. All these differences create a bell-shaped distribution, and only people with an unfortunate combination will be at one or other end of that distribution. By taking account of all the SNPs a person has relating to a particular trait, it's possible to derive a 'polygenic score', indicating the likelihood that they will become depressed, anxious, schizophrenic, obese, or much else.

With this genetic information, it now becomes possible to diagnose conditions based on their cause, rather than on the symptoms (usually a person's behaviour) – as is done now. As Plomin points out, people can be depressed for all sorts of reasons, but polygenetic scores can predict the extent to which a person will be depressed for genetic reasons.

In many ways, this creates a revolution in psychiatric treatment and in categorizing and diagnosing conditions. It also gives a way to look at what's different about people at the end of the curve furthest from the psychological condition – are they exceptionally well, or do they have their own issues associated with the genes they have?

And all these differences can be found in the 1 percent of our DNA that can be different from one person to another. The other 99 percent we all share. While, Plomin doesn't say that our environment is unimportant in making us the person we are, he emphasizes that it is not as important as our genetic predispositions. With the latest advances in DNA detection, it now becomes possible to predict not exactly what will happen in a person's life, but the likelihood of something or the degree to which that something will happen.

> Epigenetics is the study of heritable phenotype (chracteristics) changes that do not involve alterations in the DNA sequence (genotype). Epigenetics usually involves changes that affect gene activity and expression.

In May 2021, a GWAS of genetic and health records of 1.2 million people from four separate data banks identified 178 gene variants linked to major depression.

The study was led by the US Department of Veterans Affairs (VA) researchers at Yale University School of Medicine and University of California-San Diego (UCSD), and was published in May 2021 in the journal *Nature Neuroscience*.

Depression is genetically complex and is characterized by combinations of many different genetic variants. The size of the new GWAS study will help clinicians to develop polygenic risk scores to pinpoint those most at risk of developing major depression and other related psychiatric disorders such as anxiety or post-traumatic stress disorder.

References:

Robert Plomin. Blueprint: How DNA Makes Us Who We Are. 2018. ISBN-13: 978-0241282076

https://news.yale.edu/2021/05/27/roots-major-depression-revealed-all-its-genetic-complexity

How to instantly use proven techniques in trance without a script

Some power words to use with clients without using a pre-written language pattern.

Wouldn't you rather take one small step towards being able to work with clients without relying on scripts (language patterns) rather than remain unable to help a client if you don't have the right scripts with you? In fact, I know it can be easy to do. I also know that you will be so pleased when you are able to work with people anywhere at any time.

That first paragraph includes a number of sales tricks that can be used to get people to purchase expensive products, but we can also use these same techniques to help clients to achieve their goals. And, of course, these are the techniques so often used by politicians to persuade an audience!

Let's look at some of the more useful words and phrases (sometimes called power words) that we can weave into an unscripted session with a client. Here's a script to help someone sleep.

As you lie there now listening to my words, you find yourself starting to let yourself relax.

I want **you** to think about **how you'll feel when you** wake up in the morning after a good night's sleep. **Imagine** how **amazing you'll feel** having slept all through the night.

And then picture the scene when you've slept all through the night for a whole week, and you've woken up each day feeling great, so you're probably hoping that tonight will be the first of those nights.

Just imagine **what would it be like if you** slept through seven nights in a row.

Wouldn't you agree that being completely relaxed is the best way to go to sleep? And listening to my download can help **you** to achieve that.

The great thing about sleep is that it's **free** – and there's nothing **new** in the world about that.

Just let yourself **remember** for a moment, a time when **you** have been completely relaxed and so incredibly comfortable. Just feel again what it felt like. Hear again any sounds **you** could hear then – or maybe it was completely quiet. And smell and taste again any smells or tastes **you** experienced then. Let that relaxation memory just sweep over **you**.

And as **you remember** that feeling of total relaxation, **imagine what it would be like if you** could bring that feeling right back to each and every bedtime.

And **imagine**, when **you** get into bed tonight and settle down to sleep, those same feelings of relaxation sweeping over you.

And **because you** are lying still in bed, you'll be so pleased to find yourself starting to relax, and as those waves of relaxation pass over **you**, you'll **remember** that time when you were perfectly relaxed. Perhaps **you** even fell asleep then? And you can give yourself permission to fall asleep tonight. **And that means you** are already starting your week of sleeping through every night. Just **picture** it, **finding yourself** in the deepest and most tranquil state of calmness imaginable. So calm that **you realize you** can really unwind and let go. And as that feeling of serenity washes over **you, you** know that **sooner or later** you'll enjoy wonderfully peaceful relaxation and then sleep.

Just **imagine** it now. Lying in bed and naturally drifting off to sleep – **finding yourself so relaxed** that sleep just comes by itself.

As you breathe in and out, you will notice an ever-deepening comfort starting to develop, **which means** that relaxation will lead to a long and revitalizing sleep.

And if, for some reason, **you** don't find yourself drifting nicely into the welcoming arms of sleep, I want **you** to **just pretend** that you're in that most relaxing place **you** can remember. **Just pretend** that **you** are more comfortable than you've ever been before. **Just pretend** that your breathing is slowing down and helping **you** enjoy this deep and luxurious relaxation.

And **the more you** pretend that you're relaxing, **the more** comfortable **you** become. **The more** comfortable **you** become, **the more** relaxed your body feels.

And very soon, **every time you** take a breath, you'll feel more-and-more relaxed. And because of this growing feeling of relaxation inside **you**, you'll know **what it's like when** every muscle in your body is completely relaxed and completely stress free.

Just **suppose** that **you** are relaxing like that tonight. No rush, no hurry, a perfect night's sleep waiting for **you**, when **you** go to bed. And **sooner or later**, you'll **find yourself** waking up after a whole week of great night's sleep.

You can take any of the words or phrases in bold text above and sprinkle them through your own extemporized monologues during trance. It makes sense to make what you say as powerful and effective as it can be.

Let's put that to one side for the moment

A look at some useful bridging words and phrases to use.

Sometimes, clients are completely happy to follow our lead. They get the idea that our sessions are solution-focused, and they want to work on the solutions to their particular issue. However, there are some clients who think that coming for talking therapy allows them to simply talk as much as they can in the allotted time. And there are some clients who seem completely focused on their problem. It's all they've thought about for a while now and they keep coming back to it. And no matter how many times you ask them about good things (sparkling moments) in their past week, their focus is permanently on the bad stuff. I want to look at what you can do to move the conversation forward – I want to talk about bridging words and phrases.

You find bridging words and phrases being used by media-savvy politicians. They can be almost unnoticed when they are used, but they allow that politician to side-step a difficult question or statement and move onto their own agenda. And that's basically what we want to do. We want to acknowledge the client's problem. And we want to show that we have listened to their explanation of their problem. However, we don't want to find our whole session with them spent dwelling on the problem. We want to move them (and the session) forward towards a goal or solution to that problem.

> Bridging words and phrases connect two separate ideas with some sort of relationship.

The truth is, that we probably naturally do this in a conversation with a client. But, we all know there have been times when we feel like we're losing control of the session and we're wasting our time and the client's time dwelling in their past.

So, what can we say to get back control? Perhaps the phrase that we were trained to use is: "thank you for telling me that, but let's put it to one side for the moment and we may come back to it later". We've acknowledged what the client has said, but we've not given them control of the session agenda. It's a good bridging phrase.

If you're looking for a useful bridging word, then try "actually". This is a great word to use when your client has explained what someone has done and what that means in terms of their relationship or the client's social standing. You then give them a reframe – a completely different way of thinking about what happened. And you've done it in a way that has seemed perfectly natural and hasn't upset them in the least.

Other words you might use to present a different point of view include:

- (And) still
- (And) yet
- Although
- Although that may be true
- At the same time

- Be that as it may
- Even so
- Even though
- In contrast
- Nevertheless
- Nonetheless
- Notwithstanding
- Of course ..., but
- On the other hand
- Or
- Then again
- Whereas.

There are other contradictory words, but the idea is to gently lead your client towards the opposite viewpoint without making it completely obvious at the start that you hold an opposing view.

Here are some bridging phrases that you might have heard used:

- I see that, but ... (key message)
- I can't comment on that. What I would say is ...
- Can I just add that ...
- Just to put that into some context ...
- What's absolutely critical to remember is ...
- People have said that but...
- To put this in perspective ...
- That's very interesting, but first let me make the point ...
- That's very interesting, but what I believe is ...
- What we have to look at is ...

Everyone likes stories. So, sometimes, after a new client has spent more time than is necessary talking about their particular issue, you can say something like: "the reason I became a hypnotherapist was". Like I say, everyone likes a story, so your client is now focused on you. And you can tell a story about how you wanted to help people (with the client's problem) and how you found that focusing on solutions works so much better than any other technique. You've made your point and your client was following you all the way.

A bridge always works best when the client's comments are acknowledged before moving on. Here are some more examples:

- That is a concern, but what my clients tell me works for them is...

- That's an interesting point, but I think the bigger issue is...
- I'm not sure that's the case. What my experience has shown is...
- That's one point of view. What I'd also like to suggest is that...
- Of course, but to put that in perspective...
- That's very interesting, but first let me make the point...

The idea is not to sound like a politician sneakily avoiding answering an interviewer's questions – although that's where many of these bridging phrases come from – the idea is to simply get back control of the conversation from your gossipy client or your client who is focused on their problem. It's a technique that allows you to stay in control of the agenda for your session with that client and do what you do best – help them to overcome the issue that brought them to see you in the first place.

It's useful to know that you have those bridging words and phrases available to you to use to make your session with a client more like a therapy session and less like an evening down the pub with someone who is completely focused on themselves. After all, they are paying you for therapy and that's what you want to give them.

References:

https://www.smart-words.org/linking-words/transition-words.html

https://roughhousemedia.co.uk/public-relations/20-bridging-phrases-to-help-you-control-media-interviews/

https://www.mediafirst.co.uk/blog/14-bridging-phrases-for-your-next-interview/

Performance anxiety script

A script to help people with an important event ahead.

As you know ... our primitive brain thinks it's there to run the show ... it selects our favourite habits ... and that's what we tend to do ... we behave in our usual way ... we perform our usual habits ... without thinking ... unless the executive function in our intellectual brain ... the boss ... says 'no'.

And we can so easily establish new habits ... it's a bit like the actor who wore odd socks on the night they gave a stunning performance ... and, from then on ... they always wear odd socks before a show.

So ... if we rushed to the toilet three times before a half marathon ... or meeting an important person ... or giving a speech or lecture ... and the race ... or the meeting ... or the presentation ... went well ... then our protective primitive mind ... just like the actor's ... will want to always do that. It will be a new habit ... and one that the primitive brain will encourage us to do the next time ... we run a half marathon ... or meet someone important ... or give a talk.

And we get locked into the habit ... and the little thinking that the primitive brain can do ... might result in it saying that going to the toilet three time ... or whatever ... worked well ... let's try going four times! Surely that will be even better?

So, what can we do about it? What's to stop us ending up in the loo five or six times or more?

The answer is to get that intellectual brain back in control. The intellectual brain can see that ... on the whole ... our previous experiences have been positive ... and ... statistically ... our next performance is very likely to be the same.

If you were a betting person ... you can bet on person's performance being the same as their most recent performances. How many times has someone surprising been ahead on the first day of a golf tournament? Pundits say that if he continues like this, he will win by so many strokes. But ... more often than not ... that person's performance returns to average over the next couple of days ... and the competition is won by one of the better players.

I'm not saying people can't improve their performance ... far from it ... and a person's average can improve over time.

Nor am I saying that people don't have good and bad days ... what I'm suggesting is that a person's performance today will most probably be similar to their recent performances.

And if they've always been good ... then the next time they run ... or meet someone ... or give a presentation ... they will be equally good.

That primitive brain though ... will still be trying to get the hypothalamus to send messages to the adrenal glands to produce adrenalin, noradrenalin, and cortisol. And they will be getting the body into flight-or-flight mode ... which means blood will be taken from the inner body and sent to the brain and muscles ... it means that there will be little blood left around the intestines for digestion ... so your body will want to get rid of any food left in the intestines. It will want to get rid of any urine in the bladder. These will just be ... in the way.

So ... how do we get back control? How do we get our intellectual brain back in control? The answer ... simply, is to relax ... and that requires that we remove the stress from our lives ... empty our stress buckets by getting enough sleep ... and let our intellectual brains keep control.

How we respond to events depends on how we think about things ... how we interpret those events. So ... two people standing next to each other may respond to the same event quite differently ... it all depends on their view of the world. A stressful event might be laughed off by one person ... but be thought of as awful by another. And that interpretation may push them further into primitive brain thinking. We need to stay in our intellectual brains.

So, give yourself permission to relax ... empty that stress bucket ... work on positive thoughts ... positive actions ... and positive interactions ... ensure your life has purpose ... think about what happens to you ... positively ... confidently ... And your intellectual brain ... your executive function ... will be able to say 'no' to the primitive brain ... it will be able to decide whether or not you need the toilet because it will keep you calm ... in rest-and-digest mode ... there will be plenty of blood going to your intestines and bladder. You will feel relaxed ... you will feel confident and capable.

And that doesn't mean that you won't carefully plan your strategy for the race ... or what you might say to that person ... or what you're going to say in your presentation ... because you definitely will. That's what makes you so good at what you do.

And if things do go awry once in while ... you'll be able to think that's quite normal ... after all ... we are only human ... and you'll be sure that ... by the law of averages ... there's an amazingly good performance coming your way any time soon.

Relax ... stay in your intellectual brain ... prepare thoroughly ... keep the body in rest-and-digest mode ... and enjoy your undoubted success.

Put that picture at the forefront of your mind ... and live it ... more and more each day.

Some thoughts on using solution focused techniques

Some ideas on what to do with clients.

Solution-focused hypnotherapy uses techniques developed in solution-focused brief therapy (SFBT). At its centre, solution-focused therapy is never about the problem the client comes to see the therapist about, it's all about is the outcome the client is hoping to achieve – and the progress they are hoping to make.

The major tenets of SFBT are:

- If it ain't broke don't fix it
- If it works, do more of it
- If it's not working, do something different
- No problem happens all the time – there is always an exception that can be utilized
- The future is both created and negotiable
- The solution is not necessarily directly related to the problem
- Small steps can lead to big changes.

> Your beliefs become your thoughts.
> Your thoughts become your words.
> Your words become your actions.
> Your actions become your habits.
> Your habits become your values.
> Your values become your destiny.
>
> *Gandhi*

The language for solution development is different from the language needed to describe the problem.

One of the other tenets of SFBT is that the therapist believes there is a 'best' version of the client that they get to talk to. This allows the therapist to carry on, even when they may be faced with resistance, denial, and non-compliance from the client. And they carry on until these difficulties disappear. And if your clients aren't amazing people, then they will have amazing things in them that allow them to do good things. Your conversation helps them do just that.

So, solution-focused therapists always work with great clients! Others might see them as just ordinary clients, but the solution-focused therapist sees the greatness in them. Part of the therapist's job is to listen for 'greatness'. That means that SF therapists must learn to cooperate with their clients' capacities. They should explore instances of success. And co-construct their clients' identities of possibility. And they should keep in mind that there is always a solution.

The best way to achieve this is to get clients to describe their perfect future, and then help those clients to find exceptions in their current situation. The client can then see how those exceptions are part of their preferred future.

For a solution-focused therapist to be successful, they must understand the nuances of the SF approach. They must understand that a solution-focused therapist focuses on the client's desired outcome and uses only presuppositional questions – that means the therapist always presupposes that the client will achieve their goals.

For the solution-focused approach to work, the therapist must only use the solution-focused approach and not slip into any other modalities, ever! The therapist must keep believing, keep hoping, and keep going! What can make the job hard is the fact that new clients, when they first arrive, are probably at a low point in their life. In order to help those clients make changes, the therapist does need to maintain their belief in the client.

One problem many SF therapists face is that some clients will want to keep coming back to their problem. They might keep saying, "yes, but..." to every positive the therapist finds or each question (presupposing success) the therapist poses. The secret, in this situation, is for the therapist to phrase their questions in such a way that acknowledges the client's recent difficulties in the hope that the client won't need to repeat those difficulties as part of their answer. For example, the therapist might ask: "Given that it's been difficult for you recently, how have you managed to...?".

Looking at Figure 1, it's possible to see that the structure of every solution-focused session is divided into three parts. The first part is the desired outcome. This sets out what the client wants to achieve in the session or from all the sessions. Once that has been established, the main part of the session is taken up with the description – which has four parts to it. And at the end is the closing of the session. Let's look at those in more detail.

> SFBT works on the assumption that clients have an outcome that they hope to achieve, and the resources to help them work towards their preferred future.

So, the first step is to identify an outcome for the sessions – this is different from a goal. This is what the client is there to achieve. Useful questions to ask in this part are, "what is your desired outcome from this work?" Or, "what are your best hopes from this therapy?" That will identify their goal. The trouble often is that their goal won't be under their control – for example, they might say that their goal is for people to like them. The next few questions should find an outcome that the client does have full control over. You might ask, "what difference would it make to you if 'people liked you'?" Or you might ask, "suppose you woke up tomorrow and 'people liked you'. How would you feel?" Taking their answer, you might ask, "how would that make a

> When asking clients questions, every question has to be about the presence of the desired outcome and connected to the client's resources that could make that outcome a reality.

difference to your life if you were <whatever their previous answer was>?" You might then say, "suppose you woke up tomorrow and you were able to <previous answer>, and you still had <original problem>, how would you feel?"

So, they came with a problem, you got them to suggest a goal for the work – one where the problem was solved – and, through questioning, were able to find an outcome that was under their control that they could work towards. It may not solve the

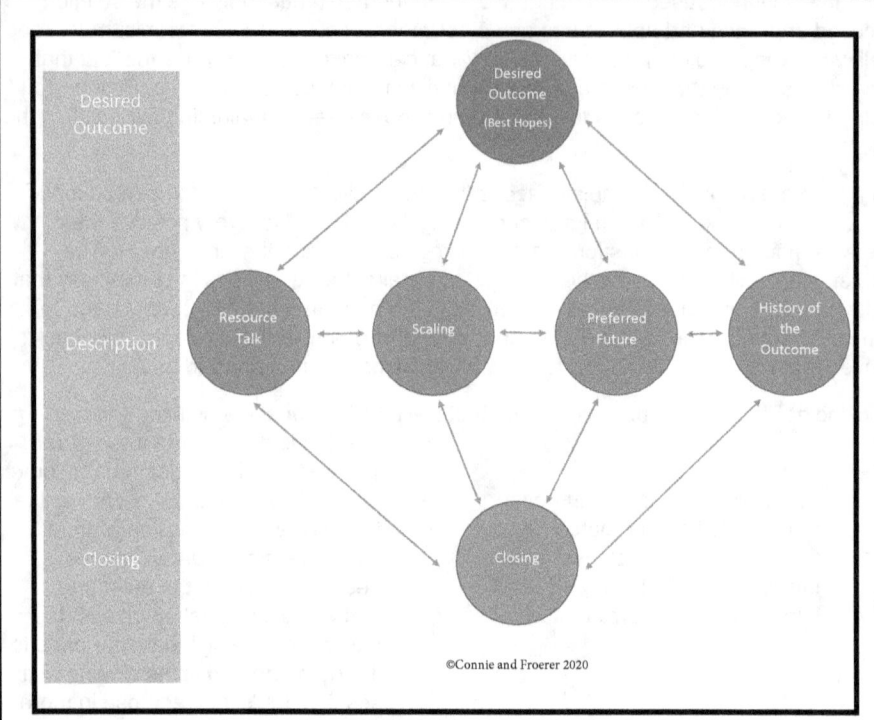

Figure 1: Connie-Froerer Diamond Model

problem as such, but it makes the problem so much less of a problem that it becomes unimportant.

The next stage is to move to a time when the client has experienced their desired outcome. The description section helps the client to see the presence of their desired outcome in their lives already. The description section has four component parts.

Resource talk highlight resources the client already has that can be utilized to accomplish change. Questions might be framed like, "wow, it's must be really hard to feel x, but if I'd met you at a time in your life when you felt the most y, what resources were you drawing on from within you to help you achieve that sense of y?" From what they say, it's possible to create a list of resources that they have used.

Scaling can be used to highlight descriptors. They will say how they feel on a scale of 0 to 10. So, a question might be framed like this: "It must be really challenging to feel x, I hope I can help in this session. On a scale of 0 to 10 where 10 represents you totally y, and 0 is the exact opposite, where would you say you are today?" They will give an answer – say n. The therapist can say: "suppose you walk out of my office and you're an n+1, what would you notice that would tell you that you're at an n+1 instead of an n?" Or: "Things must be really difficult. n seems very low, but it's not n-1. What do you notice about your life that lets you know you're not n-1?" This helps the client to identify the resource that they already have, and they are using.

If they were to say -10, the therapist can follow up by asking: "what are you doing that you're pleased with to help you cope and manage at such a difficult time in your life?"

Their **preferred future** can often result from the description they give when answering the miracle question. Other questions that the therapist can ask to get the same result are: "suppose you woke up tomorrow, even though you're having these struggles and problems, your sense of y and your sense of yourself has been returned to you, what's the very first thing that you would notice?" Follow up questions can help the client to describe the details of the discovery of having those things come back into their life.

History of the outcome is where the client talks about the presence of their desired outcome in their life previously – ie where they've achieved their goal in the past. Good questions to ask include: "in the past, when your sense of y was present in your life, how did you know it?" Or: "when these things were present, how did they make themselves aware to you?" Or: "how were you aware of the presence of these things when they were present in your life in the past?"

One problem with asking clients questions is that they may reply saying that they don't know the answer. Often, a "don't know"" answer indicates that the client has thought about this question before. It's important not to make the person feel defensive or they will never move from their "don't know" answer. You could ask them what they think the answer might be. Perhaps better is to ask them what small change a significant person in their life, or a pet, who knows them so very well, might notice. Because the person, or pet, knows them so well, they are bound to notice any change, and it is easier for the person to answer the question. And so you have successfully moved on from the "don't know" answer.

The closing part of the session is important because it provides a way to maintain the client's autonomy. They are not relying on the therapist to come up with all the answers. A good thing to say would be: "Thank you for coming, I hope it was useful. Before our next session, I want you to look for evidence or signs of change." Remember: a change is only a change if it's noticed! Once a client notices a change, they are able to achieve more changes.

One of the great 'discoveries' of de Shazer and the Milwaukee team was that 'solutions' had more in common with each other than they do with 'problems'.

One of the great techniques that comes from solution focused brief therapy is a way to help a client to be the best version of themselves that they can be. When a person has an event coming up, especially an event that will have a substantial impact on their life (eg job interview) and where their performance at the event will have a significant impact on the outcome, then 'being at their best' during the event is likely to make a difference to their future life.

The experience of 'being at their best' during an interview frees the client from a sense of responsibility for the outcome, or, more particularly, a bad outcome that might otherwise have led them to blame the misfortune on themself.

This technique is very useful for events involving strangers and containing unpredictable elements. In these cases, it is not possible for the client to describe the event in great detail because there are too many unknowns. For key events with known participants, such as a team meeting, it's possible to build an accurate 'at your best' picture of the event itself.

> For people who are stuck and can only see the problem, here's the question to ask: "What would you like to feel instead?"

Whatever the situation, the technique begins with an 'at your best' description from the start of the day, it includes the 'successful' meeting, and perhaps finishes with arriving home at the end of the day. In this way, a key but isolated event is embedded in a client's wider daily life and has available a wider array of potential resourcefulness.

When using this technique, the client is asked to describe both their inner states and the outer actions that signify these states. They are asked to locate themself at specific places and specific times, and attention is paid to the most humdrum details such as the period during which their computer turns on or while sitting on a train. They are asked to think about what this will look like to others (who may well be strangers or even pets), and how these others might respond. In the middle there is the opportunity to list some of the reasons why the client might be justified in feeling confident and how to deal with any setbacks that might be part of the vagaries of life.

The client leaves the session with no more than they came with: they are asked to describe nothing that was not already within their experience of themselves – no 'miracles', no imagining they can do something they've never done before, just simply a description of what they might ordinarily do on one of those days when they are at their best.

However, what makes this technique so special is the extraordinariness of the questions and therefore the never-before-spoken answers. The client is unlikely to have been asked what a stranger might notice about them, so their answer is likely to be one that they have never heard themself say before that moment. At least half of their answers will also be new to them, and, many of the rest, though not necessarily new, will have new meanings. Though almost all the content comes from the client, the therapist, by the way they choose to follow each of the client's answers, is very much in charge of the direction the session takes.

What is completely missing from this conversation is the therapist's own experience of a similar event. The therapist must stop themselves passing on the benefit of their own experience – which can be very difficult!

Brain food

A look at some of the foods that are good for your brain, and some that aren't!

Wouldn't it be great if we could recommend foods that were good for our clients (and our) brains, and suggest to them to avoid some foods that are harmful.

The problem is that there are so many different diets advertised that people are confused about what's good and what's bad. Plus, people have their favourite foods and just like to eat those – no matter whether they are healthy or not. And there seems to be an issue with portion size. These days, many people are used to 'going large' with what they eat and making changes to their diet is just too hard.

So, I'm not telling anyone what they should eat and what they shouldn't eat, I'm simply looking at the evidence from research and letting clients make up their own mind. And, because any diet should be varied, if you eat from the 'naughty' list every now and again because that's what you fancy eating or because you don't have time for anything else, then that's OK. People are more likely to stick to a diet that allows variety than one that is restrictive.

A good place to start with diet is to see what people who live long healthy lives eat – the people who don't suffer from Alzheimer's or similar conditions. It would make sense to see whether what they eat matches the research into healthy food. Perhaps, the best-known diet is the Mediterranean diet. This comes with a lower risk of developing diabetes, obesity, cardiovascular disease, and Alzheimer's. People on this diet eat a wide range of vegetables, fruits, beans, and nuts. They eat a variety of whole grains like wheat, oats, spelt, and barley. And they eat wild-caught fish. Their food comes with lots of herbs and spices. And they use lots of extra virgin olive oil, and drink red wine with their meals. They do eat meat and dairy, but not very often. And they only occasionally eat sweet puddings and in small portions. A meal is usually a social event.

Another region famed for longevity is in China. It's called Bama Yao in Guangxi. It's a poor region, so people eat frugally. They eat mainly fresh vegetables and fruit. They also eat rice, maize kernels, sweet potatoes, nuts, seeds, maize, beans, peas, lentils, and freshly-caught fish. They use a lot of hemp-seed oil. Meals are often social, and elders (who are revered) are served first. They also use the herbs ginko and ginseng.

So, that would be a good place to start. One day eat a Mediterranean diet and another day the Chinese diet of the people from Bana Yao.

Our brain can only get the nutrition it needs from the food we eat – that's pretty obvious. But does what we eat have any obvious impact on the brain? The answer is yes. Frighteningly, brain scans of similarly-aged people who eat either a standard western diet or a Mediterranean diet show that the brain of the person on the western diet looks older. The brain has shrunk (atrophied) in certain areas. So, maybe, fizzy drinks, fast food, processed food, and refined sugary sweets aren't that good for you. Not if you want to enjoy a healthy old age.

Let's look at human diet in evolutionary terms. Up until around 1.8 million years ago, the fossil record shows that our human ancestors' brains were about the size of a chimp's. After that time, the brains got bigger (Homo erectus). And bigger brains need more energy from food (and oxygen from the air). It would need far more food than just eating the low energy plants around them could provide. It used to be thought that these early hominids must have caught and eaten meat. But because of the size of these hominids and the tools available to them, it looks unlikely that they could have caught meat. So, what's the alternative explanation? These early hominids lived near water and evidence suggests that they caught and ate fish. Fish and shellfish would have provided the energy-dense brain food that they needed (along with the fruits and vegetables that were around them). The next big leap in brain evolution came with fire. Cooking produces soft food (while retaining the calories), which took less time to chew and digest. Humans

Fats are made of long chains of carbon atoms. Some carbon atoms are linked by single bonds and others are linked by double bonds. A saturated fat is a type of fat in which the fatty acid chains have all or predominantly single bonds.

An unsaturated fat is a type of fat in which there is at least one double bond within the fatty acid chain. It is polyunsaturated if it contains more than one double bond.

with smaller teeth and a shorter GI tract, plus a bigger brain, was the result. Then came livestock and then farming, and then civilization as we know it! Homo sapiens had big brains and ways of keeping food available at all times (except in catastrophic situations). The human population just grew and grew.

Omega-3s come in three types: alphalinoleic acid (ALAs), eicosapentaenoic acid (EPAs), and docosohexaenoic acid (DHAs).

That would suggest that the foods our bodies are designed to eat are vegetables, fruits, nuts, and seeds, as well as fish.

I haven't mentioned the most important nutrient for the brain, and that's water. A large percentage of the brain is water. Water contains minerals that are used by the body. So, drink water or coconut water or aloe vera juice.

About 11 percent of the brain is fat, and a similar amount is protein. The brain can make most of the fats that it uses. Essential fatty acids (EFAs) are

Trans fats are artificially created by hydrogenating (adding hydrogen atoms) to unsaturated fats. They increase the shelf life and flavour-stability of foods.

Eating trans fats increases the risk of coronary artery disease in part by raising the levels of low-density lipoprotein (LDL – 'bad cholesterol'), lowering levels of high-density lipoprotein (HDL – 'good cholesterol'), increasing triglycerides in the bloodstream, and promoting systemic inflammation.

the ones that the body needs to eat because it can't make them. The best fats to eat for your brain are polyunsaturated fats (PUFAs). The only saturated fats the brain needs are butyric acid (found in milk), and myristic acid (found in coconut

The glycaemic index (GI) is a rating system for foods containing carbohydrates. It shows how quickly each food affects your blood sugar (glucose) level when that food is eaten on its own. Foods that break down quickly in the body and cause a rapid increase in blood glucose (like sugary drinks) have a high GI rating. Low GI foods, such as wholegrains, fruit, vegetables, beans, and lentils, are preferable.

oil). The brain needs omega-3s and omega 6s, which are examples of PUFAs. The best source of omega-3 is black caviar, but salmon is quite good. Flax seeds and hemp seeds are also good. A number of studies have shown that a diet high in saturated fat is bad for you cognitively. Saturated fats are found in butter, lard, fatty meats, and cheese. Trans fats are the worst types of fat because they are linked with cognitive decline and dementia in old age. They are usually found in processed foods.

Oxidative stress occurs when there is an imbalance of free radicals and antioxidants in the body. Cells in the body produce free radicals during normal metabolic processes. Free radicals are atoms that have unpaired electrons in their outer shell, which makes them very reactive. Usually, the body produces antioxidants, which are chemicals that react with free radicals, neutralizing them. The body's natural immune response can also trigger oxidative stress, which causes mild inflammation that goes away after the immune system fights off an infection or repairs an injury.

Interestingly, the brain makes its own cholesterol. Worryingly, it's been found that high levels of blood cholesterol in midlife puts a person at risk of developing dementia in later life.

Proteins are composed of amino acids, and digestion breaks down proteins into amino acids. Amino acids are used in neurotransmitters. Essential amino acids are the ones that the body can't make itself. They are tryptophan (a precursor of serotonin), methionine, leucine, isoleucine, lysine, and histidine. They are found in fish, milk, eggs, and meat, as well as legumes and grains.

Carbohydrates are the third big food group. They are digested into simple sugars, which can be absorbed. The brain uses only glucose for energy – so a constant supply is needed for brain function. In the event of the body's glucose level running low (hypoglycaemia), the body will become unconscious. The best source of glucose is honey. The average brain burns 62 grams of glucose per day – about 250 calories.

Like with diabetes, the hippocampus can experience insulin resistance, which can lead to inflammation, which can make it very hard to remember anything. So, too much sugary food can be bad for your brain.

Fibre comes in two types. Soluble fibre turns into a gel-like texture when eaten and slows down your digestion, making you feel fuller longer. Insoluble fibre adds bulk to your stool, which helps the gut eliminate waste more quickly. Low-fibre diets are bad for your gut. There's lots of evidence linking the gut biome (the bacteria etc that live in the gut) with a person's mood.

The brain suffers from oxidative stress because of all the thinking it does. So, it needs antioxidants from your diet. Here are some antioxidants and their sources:

- Vitamin A from dairy produce, eggs, and liver.
- Vitamin C from most fruits and vegetables, especially berries, oranges, and sweet peppers.
- Vitamin E from nuts and seeds, sunflower and other vegetable oils, and green, leafy vegetables.
- Beta-carotene from brightly coloured fruits and vegetables, such as carrots, peas, spinach, and mangoes.
- Lycopene from pink and red fruits and vegetables, including tomatoes and watermelon.
- Lutein from green, leafy vegetables, papaya, and oranges.
- Selenium from rice, wheat, and other whole grains, as well as nuts, eggs, cheese, and legumes.

And, it's important to get the necessary mineral salts as well as the vitamins the body needs.

So, what foods should you avoid, for the sake of your brain? The easy answer seems to be anything processed or any fast foods.

What foods should you eat less of? The answer seems to be meat and puddings. And, perhaps, everything! There is a lot of evidence that eating smaller portions is good for your brain and fasting can be good for you – but check with your doctor.

And what should you eat more of? Fish – say three times a week. Drink plenty of water. And make sure the food you eat is free range (if it's an animal) and organic.

It's worth noting that the Dirty Dozen are the twelve foods that are most contaminated and pesticide laden (unless organic). They are: apples, celery, cherries, grapes, kale, nectarines, peaches, pears, potatoes, spinach, strawberries, and tomatoes. The Clean 15 are foods safe to eat in their non-organic form. They are: asparagus, aubergine, avocados, broccoli, cabbage, cauliflower, honeydew melon, kiwi fruit, maize (sweet corn), melon (cantaloupe), mushrooms, onions, papaya, peas (frozen), and pineapple.

The smoke point of oil is the temperature at which it stops shimmering and starts smoking. Smoking is a sign that the oil is breaking down. This can release chemicals that give food a burnt or bitter flavour, as well as releasing free radical. It's important, when using any oil, to make sure that its smoke point is below the temperature you plan to cook at. Extra Virgin Olive Oil's smoke point is 163-190°C. Coconut oil is 177°C, but refined coconut oil is 232°C. Lard is 188°C. Rapeseed oil is 204°C. Refined Avocado Oil is 270°C. Deep frying food is usually cooked between 177°C and 191°C. Pan frying ranges from 160°C to 205°C.

Will you get fat? Christakis and Fowler (2007) published in the *New England Journal of Medicine* a study of 12,067 people from 1971 to 2003. Worryingly, they found that a person's chances of becoming obese increased by 57% if they had a friend who became obese. They concluded that obesity appears to spread through social ties!

So, when you start to plan your menu for the next week, there are some things that you want to eat more of and some less. Here's the list:

- Water or herbal teas – every day.
- Low GI fruit, eg berries, cherries, citrus, kiwi, apples, and leafy greens and/or cruciferous vegetables (broccoli, cauliflower, cabbage, etc) – every day.
- Medium GI (plums and peaches) and/or high-fat fruit (avocado and olives) or all other vegetables – every day.
- Whole grains or legumes (beans and peas) or sweet potatoes – every day.
- Yoghurt or fermented foods – every day.
- Fish or shellfish – three times a week.
- Nuts or seeds – three times a week.
- Eggs – once a week.
- Poultry or organic cheese – twice a week.
- Red meat – no more than once.
- 1oz dark chocolate – five to seven times a week.

It's your client's choice. They can continue eating large amounts of takeaways, processed foods, and foods high in sugar, but they need to be aware of the consequences on the brain and body. Or they can start to move their diet towards that of people who live to a healthy old age. These are some suggestions of what's good for their brain. It's not an exclusion diet, and usually knowing that you can eat something reduces the drive to actually eat it. There are plenty of recipes at https://www.lisamosconi.com/beverages.

References:

Dr Lisa Mosconi. Brain Food: How to Eat Smart and Sharpen Your Mind Paperback. Penguin Life (17 Jan 2019). ISBN-10: 0241381770

https://www.nejm.org/doi/full/10.1056/nejmsa066082

Psychological First Aid

Some thoughts on what to do in an emergency.

We're all familiar with people coming to see us to help them with their particular problems. In the comfort of our own consulting rooms or on Zoom, it can be easy to tease out the facts, ask the usual useful solution-focused questions about what strengths they use when the problem doesn't happen or when is it less severe, and we can formulate, in our minds, the best way to help them. And, working together, we do help. We sometimes refer to ourselves as psychotherapists. And we offer psychological help with our trance work.

But what about if we come across someone struggling with a panic attack? Or what if we find ourselves at the scene of a major accident or terrorist attack or some other catastrophe. In the latter case, there will be people hurrying around to offer first aid to the victims, but who will be offering psychological first aid (PFA)? Who will be empowered to help survivors or casualties, or even simple observers, deal with the emotional stress etc that they will be going through? Who better than solution-focused hypnotherapists – with our knowledge of the brain and our experience of helping people?

First aiders are trained to look for quiet people who may not be breathing, or people bleeding profusely, and work their way from more life-threatening casualties to more noisy people, who are able to breathe and call out. They have rehearsed dealing with this kind of situation. But, quite often, casualties are slow to recover because of the overwhelming stress they have experienced. So, what could we do to help? Here are some thoughts about psychological first aid.

Once people are no longer afraid for their lives, they often have to cope with negative emotions such as anger, anxiety, and hopelessness. They may also experience stress-related difficulties with memory and decision making. What we can do is help people understand and cope with their distress, which makes it easier for them to begin the recovery process.

So, what exactly is psychological first aid?

The World Health Organization in 2010 defined psychological first aid as: humane, supportive response to a fellow human being who is suffering and who may need support. It entails basic, non-intrusive pragmatic care with a focus on listening but not forcing talk, assessing needs and concerns, ensuring that basic needs are met, encouraging social support from significant others, and protecting from further harm.

According to George S Everly Jr, PFA may be defined as a supportive and compassionate presence designed to do three things:

- Stabilize (prevent the stress from worsening)
- Mitigate (de-escalate and dampen) acute distress
- Facilitate access to continued supportive care, if necessary.

According to the World Health Organization (WHO), PFA involves the following themes:

- Providing practical care and support that does not intrude
- Assessing needs and concerns
- Helping people to address basic needs (for example, food and water, information)
- Listening to people, but not pressuring them to talk
- Comforting people and helping them to feel calm
- Helping people connect to information, services, and social supports
- Protecting people from further harm.

Others suggest that successful psychological first aid has five components. They are:

- Helping people feel safe. This can be thought of as the bottom level of Maslow's hierarchy of needs. People need to feel safe. They may need food or have other basic needs that have to be satisfied. Once those basic needs are satisfied, people can cope better.
- Creating a sense of calm. Helping people manage their own emotions can be achieved by listening to them while they talk (or cry or complain). Then help them to reframe the situation so they can begin to think about how to move forward.
- Helping people to regain a sense of control and self-efficacy. People may well be struggling with feelings of helplessness. Engage them in problem-solving and allow them to determine what they need to cope. This can be empowering. Asking people to help in some way can give them feelings of agency and really help them.
- Social connection. People may feel isolated and alone. It can be helpful to engage them in an interaction with other people.
- Hope. Humans have survived all sorts of disasters in their history. Encourage people to be hopeful of the outcome.

Of course, faced with a situation, how do you get yourself to actually start? One simple technique is called the 5-second trick. You simply count down 5-4-3-2-1-go. And with that, you get up and go and start helping people.

But it's not just natural disaster or car crashes etc, where people could benefit from PFA. It could be friend who asks for help, or a person telling us about a mutual friend who could benefit from PFA (although they probably won't refer to it as that). Or it might even be someone you see around you who looks like they might benefit from some help. So, how do you actually begin to engage with them? George S Everly Jr suggests three ways that PFA might begin:

1 A person may contact you and ask for assistance. Perhaps, a friend calls saying, "I'm feeling really stressed, do you have some time for a chat?" In this case, respond with something like, "Yes, what's going on?" In many instances, the person simply wants to vent. They don't have any expectation that you will resolve the

issue at hand. In these situations, it's important to listen. Don't rush to diminish or solve the problem, unless that is the expectation. Once you've heard what they have to say, try saying something like, "I'm sorry you are going through that, how can I help?" This sets things up to help the person make a plan to address the problem (if possible), or to develop a plan to help them better cope with an issue that might be on-going. If the person turns down your assistance, you could say, "Well let me know if it turns out there is something I can do, even if it's just listening". Follow up with that person in a day or two to see how they are doing.

2   Someone might approach you on someone else's behalf. They will suggest that you speak to the third-party because something seems wrong. Be sure to clarify specifically what makes them believe something is wrong with the third party. Perhaps your child's teacher contacts you saying your child seems to be having a hard time at school. Your response might be, "What specifically is concerning you?" Once you have all the information, and at the right time, you could say to your child, "Your teacher spoke to me yesterday. They said you seem to be xxx lately. What's going on?" After listening to what your child has to say, it might be useful to say something like, "I'm sorry you are going through that, how can I help?" If there is really nothing you can do to assist directly at that point, then just say, "Well let me know if it turns out there is something I can do, even if it's just listening". Again, following up in a day or two is important.

3   Someone might look or sound distressed. Based on your concern for that person's wellbeing, you approach them, but what do you say? How about, "I couldn't help but notice you don't seem yourself today." Or, "I couldn't help but notice you seem distressed." If they respond, ask, "How can I help?" Then help them develop a plan to address the problem, or help them develop a plan for coping, or both. If they do not respond, or say there is nothing you can do, simply say "I'm a good listener." Again, if possible, follow up with that person at an appropriate time to see how the person is doing.

What else can you say to people who look like PFA might help? Here are some ideas:

- "I'm here for you"
- "I'm sorry you're not feeling well"
- "How can I help you?"
- "You don't have to go through it alone"
- "I cannot even begin to imagine what you are going through. But I will try to understand it as best I can."
- "It's not your fault".

The question that most people ask is what should they actually do in a crisis situation? Here are some positive ideas to bear in mind:

- Try to find a quiet place to talk, and minimize outside distractions.
- Respect privacy and keep the person's story confidential, if this is appropriate.

- Remain calm when speaking to a person in distress. Show concern but be a confident reassuring presence. The other person will gain confidence from your confidence.
- Stay near the person but keep an appropriate distance depending on their age, gender, and culture.
- Listen. Encourage the person to talk about what happened and their reactions to those events. If the person does not want to speak at that time, ask if you can check back with them later.
- Let them know you are listening; for example, nod your head or say "hmmmm...."
- Be patient and calm.
- Provide factual information, if you have it. Be honest about what you know and don't know. You can say, "I don't know, but I will try to find out about that for you".
- Give information in a way the person can understand – keep it simple.
- Acknowledge how they are feeling and any losses or important events they tell you about, such as loss of their home or death of a loved one. You might say, "I'm so sorry. I can imagine this is very sad for you."
- Acknowledge the person's strengths and how they have helped themselves.
- Allow for silence.
- Try to identify "the worst part" of the situation, if possible. Do this carefully, and it can help you identify the core issues at hand if they are not otherwise obvious.
- Try to determine what else, if anything, is needed after your initial conversation. Don't hesitate to ask for guidance or assistance from a healthcare professional if you are worried about the person's wellbeing.
- Serve as a liaison to connect the person with continued assistance, if necessary.
- Advocate for this person in seeking further assistance, if necessary.
- Where possible, follow up a day or so later to see how the person is doing.

And, of course, there are some things to avoid doing:

- Don't get caught up in the situation. Remember the antidote for stress is calmness and confidence.
- Don't pressure someone to tell their story.
- Don't interrupt or rush someone's story (for example, don't look at your watch or speak too rapidly). However, do interrupt if the disclosure seems to be escalating the distress.
- Don't touch the person if you're not sure it is appropriate to do so.
- Don't judge what they have or haven't done, or how they are feeling. Don't say: "You shouldn't feel that way," or "You should feel lucky you survived".
- Don't be dismissive. Don't minimize their concerns or say, "Well at least..." as an attempt to distract, or help the person feel better.

- Don't make up things you don't know.
- Don't use terms that are too technical.
- Don't tell them someone else's story.
- Don't talk about your own troubles.
- Don't make promises you can't keep and don't give false reassurances.
- Don't think and act as if you must solve all the person's problems for them.
- Don't take away the person's strength and sense of being able to care for themselves.
- Don't act on some preconceived notion of what you think the person needs. Ask what they need (Everly, Brelesky & Everly, 2018). Perspective taking such as this will foster trust.
- Don't talk about people in negative terms (for example, don't call them "crazy" or "mad").
- Don't rush. If the person is medically stable and safe, the passage of time alone begins to de-escalate situations.
- Don't hesitate to ask specific questions about the person's ability to competently attend to others (significant relationships, childcare, eldercare) or perform the duties of their job.
- Don't hesitate to ask about any intention to harm themselves or others. Seldom will this be an issue, but sometimes you may sense feelings of profound hopelessness, depression, anger, or vindictiveness. In such cases, it's important to inquire and follow up. In the most rare and extreme cases, you may have to help the person get immediate professional care.

Let's look at a real-life example: you're at the airport and someone seems to be having a panic attack. What can you do? Firstly, how do you know it is a panic attack?

According to the NHS website, a panic attack can be very frightening and distressing, and the symptoms include:

- A racing heartbeat
- Feeling faint
- Sweating
- Nausea
- Chest pain
- Shortness of breath
- Trembling
- Hot flushes
- Chills
- Shaky limbs

- A choking sensation
- Dizziness
- Numbness or pins and needles
- Dry mouth
- A need to go to the toilet
- Ringing in their ears
- A feeling of dread or a fear of dying
- A churning stomach
- Tingling in their fingers
- Feeling like they're not connected to their body.

Panic attacks typically last between 5 and 20 minutes, although some have been reported to last up to an hour.

So, how can you help someone? Here are some techniques:

- Get them to breathe slowly in through their nose for a count of four and out through their mouth for a count of eight. Do this several times.
- Distract them in some way. For example, ask them to count backwards from 3,000 in sixes. Or ask them to count all the 'T's on a webpage. Or look at a poster and count the colours or shapes. You can try anything that will distract them.
- Talk to them and reassure them that they are safe, and they will be OK. During the panic attack, they will be misinterpreting their situation as dangerous.
- Ensure they are grounded into the here and now. To that end, ask them what the date is or what the time is. Ask them to tell you five things they can see, five things they can hear, five things they can touch, five things they can smell.

If they are concerned that a panic attack may come on (or come on again) you can help them to soothe themself. They can do that by listening to music, sucking on a sweet, holding an object or photo that's important to them and focusing on that. This can help to keep them grounded and prevent the panic attack coming on.

Hopefully, armed with this information, you'll feel more confident in helping people with psychological first aid, should you find yourself in a situation where this is needed. There is one caveat to bear in mind. Although PFA does not entail diagnosis or treatment, like physical first aid, it requires basic training to be effective and reduce the risk of inadvertently making things worse (Everly & Laling, 2017).

References:

https://www.psychologytoday.com/us/blog/mental-health-matters/202007/we-could-all-use-some-psychological-first-aid

Urs Wegerhoff and Eric Zarth. Psychological First Aid: tips, tricks and techniques how to get through emergency situations. ISBN-13: 978-1791884086

https://en.wikipedia.org/wiki/Psychological_first_aid

https://www.psychologytoday.com/gb/blog/when-disaster-strikes-inside-disaster-psychology/201810/psychological-first-aid

Psychological first aid: Guide for field workers. WHO. https://www.who.int/mental_health/publications/guide_field_workers/en/

https://www.nhs.uk/conditions/panic-disorder/

https://www.inverse.com/mind-body/panic-attack-signs-what-to-do-if-you-have-one

Did you ever wonder what an ideomotor response was?

Ideomotor responses, what they are and how they are used.

Like me, you've probably read and heard about ideomotor responses (IMR), but were never quite sure what they were. Basically, they are movements made unconsciously by a client. A thought or a mental image causes a muscle movement to happen automatically or reflexively, without the person being aware of it. And, often, the movements are very small, and the client may not even be aware them.

These are quite different from the movements we ask clients to make during a rewind session, which are conscious and need to be obvious.

They are sometimes referred to as ideomotor phenomena or ideomotor reflex. Ideomotor, ('ideo' meaning 'idea', and 'motor' meaning 'muscular action') was first used in 1852 by William Carpenter in a paper about spiritualism. They referred to muscular movement outside of a person's consciousness. Carpenter was a friend of James Braid, who is sometimes referred to as the founder of modern hypnotism, and he took up the idea.

Some hypnotherapists use ideomotor responses to communicate directly with a client's unconscious mind and bypass their conscious mind. The client will use finger movements to indicate answers to questions, such as 'yes', 'no', 'don't know', and 'don't want to answer'.

Some hypnotherapists use ideosensory responses, which is where a client experiences a sensation rather than a makes a movement. So, a feeling of pain can get worse for 'yes' and reduce for 'no'.

The therapist might say to a client: "every time that I ask a question that your subconscious mind knows the answer is yes – even if your conscious mind doesn't agree – I want you to twitch the index finger on your right hand as if a fly had just landed on it. And that will signal to me that the answer is yes." The index finger on the left hand can be 'no' or no movement at all can be taken for 'no'. And, usually, the therapist will want a sign from the unconscious that it is happy to continue – so the index finger on the right hand should move. The problem with this can be that the movements are quite small, and the therapist needs to wait long enough for the signal to be made by the unconscious because there can be quite a delay. The delay reduces as the client gets used to the process. Some therapists get clients to practice raising their index finger if they are not raising it as an IMR.

For hypnotherapists who use regression, IMR can be used to step clients back through their life to find the age at which the event that is important to the client's treatment happened. When the therapist says the appropriate age, the client moves their finger. Then the problem (habit) can be dealt with by the therapist, if they are trained to work that way.

It's said that sometimes the unconscious mind will reveal information that the conscious mind doesn't know or remember.

Milton Erickson used IMR. Many hypnotherapists swear by it. Some use it as an alternative to regression and parts therapy. I'm not sure that I will be using it with clients any time soon – I'm certainly not looking for small movements of a client's index finger over Zoom!

References:

https://en.wikipedia.org/wiki/Ideomotor_phenomenon

https://www.nlplifetraining.com/general-articles-ideomotor-response-unconscious-signal-convincers

https://www.hypnoticworld.com/strategies/ideomotor

https://britishhypnosisresearch.com/ideo-motor-signalling-hypnosis-technique/

More Hypno Wiki

Active listening   This technique requires the listener to fully concentrate, understand, respond, and then remember what is being said.

Allostatic load   This is 'the wear and tear on the body' that accumulates as an individual is exposed to repeated or chronic stress. The term was coined by Bruce McEwen and BS Stellar in 1993.

Breathing techniques   Breathing deeply is a stress reliever that has many benefits for the body – oxygenating the blood, relaxing muscles, and settling the mind. Breathing exercises can be carried out anywhere.

This is an example. Breathe out completely, and then, as you breathe in, sense what's happening with your side, chest, and abdomen. Then breathe out and sense what's happening to your body. You can repeat this about ten times and feel yourself becoming more relaxed each time.

Cytokines   These are a category of small proteins that are important in cell signalling.

Empathy   Empathy is about understanding another person's situation from their point of view. It's putting yourself in someone else's shoes. It's when you understand the feelings of another but do not necessarily share them. It involves listening to their experience. It's all about them. Showing empathy means understanding what's happening to someone from their point of view. In fact, even just trying to understand can help. Don't judge or make assumptions about their experiences, thoughts, or feelings. Do recognize the emotions the other person is feeling, and show this by reflecting back what they're saying, eg say, "that sounds really difficult because...". It keeps the focus on them. Empathy helps to develop a sense of connection which is important if someone feels isolated or alone.

Grey matter   This can be found at the surface of the cerebral hemispheres (cerebral cortex) and the cerebellum, as well as in the depths of the brain (thalamus, hypothalamus, basal ganglia, etc), and the brainstem.

Guided imagery   Because people find it easier to focus on 'something' rather than 'nothing', guided imagery can be an easier technique for many than meditation. CDs of natural sounds can be played as a way of promoting a more immersive experience. Tropical Island is a good guided-imagery script.

Thirty seconds is enough time to shift a person's heart's rhythm from stressed to relaxed, It will also help to slow a person's breathing, relax tense muscles, and put a smile on their face! Creating a positive emotional attitude also calms and steadies a person's heart rhythm, contributing to feelings of relaxation and peace.

Hypnosis	This involves focused attention, reduced peripheral awareness, and an enhanced capacity to respond to suggestion. Altered state theories see hypnosis as an altered state of mind or trance, marked by a level of awareness different from the ordinary state of consciousness. Nonstate theories see hypnosis as, variously, a type of placebo effect, a redefinition of an interaction with a therapist, or a form of imaginative role enactment.
Hypnotherapy	This is using hypnosis for therapeutic purposes. It incorporates some of the features of guided imagery and visualizations, with the added benefit of enabling the therapist to communicate directly with a person's subconscious mind to enhance their abilities, more easily give up bad habits, feel less pain, more effectively develop healthier habits, and help raise their self-esteem and confidence. It may use direct and indirect suggestions.
Inflammation (update)	An initial response of the immune system to infection. Symptoms include: redness, swelling, heat, and pain. These are caused by increased blood flow into tissue. There is evidence linking inflammation with depression.
Meditation	Meditation releases DHEA (dehydroepiandrosterone) and melatonin – two useful hormones for healthy bodies. In addition, mentally focusing on nothing helps a person to not worry about other issues going on in their life, which could otherwise increase their cortisol level. Any repetitive action can be a source of meditation, including walking, swimming, painting, knitting. A person can be meditating while carrying out any activity that helps keep their attention calmly in the present. The Loving-kindness meditation is an example, and it can help people to feel better and then behave in a more positive way. Christian meditation is different from Eastern meditation.
Oxidative stress	This refers to an imbalance of free radicals and antioxidants in the body, which can lead to cell and tissue damage. The body's natural immune response can trigger oxidative stress temporarily. This type of oxidative stress causes mild inflammation that goes away after the immune system fights off an infection or repairs an injury.
Oxytocin	Oxytocin (aka the cuddle hormone or the love hormone) is produced by the pituitary gland. People with high levels of oxytocin have much lower levels of oxidative stress.
Progressive muscle relaxation	By tensing and relaxing all the muscle groups in your body, you can relieve tension and feel much more relaxed. Start by tensing all the muscles in your face for ten seconds. Then completely relax for ten seconds. Repeat this with your neck,

followed by your shoulders, etc. Tensing your muscles also boosts your willpower (see *Rip It Up: The radically new approach to changing your life* by Prof Richard Wiseman).

Relaxation
: A relaxed person has an absence of arousal that could come from sources such as anger, anxiety, or fear. Their body and mind are free from tension and anxiety. Relaxation helps a person to cope with stress. Relaxation techniques include breathing techniques, meditation, guided imagery, and other methods.

Sympathy
: Sympathy is more about how you feel. It's when you share the feelings of another. And it's associated with feelings of pity or sorrow. Sympathy can be more about managing your own feelings than understanding someone else's feelings, and is often expressed as making someone feel better or trying to cheer them up. This, unfortunately, can create more of a sense of disconnection. Saying to someone that they'll feel better soon or describing when something similar happened to you could make the other person feel as if you don't want to hear what they're going through, even though it might make you feel like you've helped.

Vagal tone
: The health and fitness of the vagus nerve is called vagal tone. A high vagal tone equates to a better capacity to keep inflammation down. Vagal tone can be improved by thinking kind thoughts about other people.

Visualizations
: Building on guided imagery, a person can imagine themself achieving goals like becoming healthier and more relaxed, doing well at tasks, giving a successful presentation, etc. Remember, visualizing doing a task is as good as physically doing the task in terms of improvement. It's the well-known basketball team experiment (http://www.breakthroughbasketball.com/mental/visualization.html).

Try creating a peaceful visualization, or 'dreamscape'. To start, simply visualize anything that keeps your thoughts away from things that are currently worrying you. It could be a favourite holiday spot, a fantasy island, country walk, or something 'touchable', like the feel of silk etc. The idea is to take your mind off your stress, and replace it with an image that evokes a sense of calm. The more realistic your daydream, in terms of colours, sights, sounds, smell, and touch, the more relaxation you'll experience. And once you're relaxed, you can imagine doing or dealing with something you find challenging.

Guided imagery simply helps you to relax; visualizations help you to relax and picture yourself successfully achieving a goal.

To achieve a goal, it is also important to visualize the steps that need to be taken.

White matter
: This is the name given to bundles of myelinated axons that connect different grey matter structures throughout the entire brain.

About the author

Wellbeing and mindset specialist, Trevor Eddolls BA, Cert Ed, MOS MI, DHP, HPD, SFBT Sup (Hyp), CBT (Hyp), Dip Hyp (paediatrics), Dip Mindfulness, AfSFH (Exec), CNHC Registered, is a clinical hypnotherapist and psychotherapist. He's clinical director at iTech-Ed Hypnotherapy and Head of IT and Social Media on the AfSFH (Association for Solution-Focused Hypnotherapy) Executive. Trevor is a Hypnotherapy Master Practitioner, a Solution-Focused Hypnotherapy Supervisor, and an NLP Master Practitioner. He is a qualified Life Coach, He also has diplomas in Positive Psychology, Counselling, Nutrition, and Play Therapy.

Solution-focused hypnotherapy, as its name suggests, focuses a client's attention on the solution to their problems rather than the causes. Evidence suggests that dwelling on what led to a problem can increase the client's issues, whereas focusing on solutions can dramatically reduce those issues.

Trevor is a popular blogger, presenter, and podcaster. He has been seeing clients and writing about hypnotherapy, CBT (Cognitive Behavioural Therapy), NLP (Neuro-Linguistic Programming), Mindfulness, and Positive Psychology techniques for over 10 years.

He also runs the hypnotherapy training company, SFHPlus. And runs regular CPD (Continuous Professional Development) courses for hypnotherapists.

Before training as a hypnotherapist, Trevor worked with mainframes. He also spent many years writing books and articles, and editing well-respected technical journals about mainframe technology.

You can contact Trevor at iTech-Ed Hypnotherapy in the Wiltshire town of Chippenham.

His websites are at https://.ihypno.biz and https://sfhplus.co.uk

Facebook: facebook.com/iHypno2004

Twitter: twitter.com/iHypno2004

Instagram: instagram.com/ihypno2004

Podcasts: https://rss.com/podcasts/solutions/

www.ingramcontent.com/pod-product-compliance
Lightning Source LLC
Chambersburg PA
CBHW070432290526
45791CB00005B/1942